FORENSIC USES of DIGITAL IMAGING

John C. Russ

CRC Press
Boca Raton   London   New York   Washington, D.C.

Library of Congress Cataloging-in-Publication Data

Russ, John C.
  Forensic uses of digital imaging / John C. Russ.
    p. cm.
  Includes bibliographical references and index.
  ISBN 0-8493-0903-4 (alk. paper)
  1. Forensic engineering. 2. Image process—Digital techniques. I. Title.

TA219.R87 2001
621.36′7—dc21   2001016190

No claim to original U.S. Government works
International Standard Book Number 0-8493-0903-4
Library of Congress Card Number 2001016190
Printed in the United States of America 1 2 3 4 5 6 7 8 9 0
Printed on acid-free paper

Introduction

The need for a book dealing specifically with the forensic uses of digital imaging was made plain to me a few years ago, after four days of testimony as an expert witness in a murder trial, in which the guilt or innocence of the defendant hinged in large measure on whether or not he was the person in the images on a poor quality and rather vague videotape from a surveillance camera. Interpretation and misinterpretation of information about imaging in general and digital image manipulation in computers in particular by expert witnesses on both sides, and by counsel who asked questions that were confused and confusing, may or may not have really helped the jury in reaching a decision. They convinced me, as a participant, that the information that can be found in a variety of standard reference and textbooks is not adequately accessible to witnesses and counsel.

Experiences as an expert witness in other trials, both criminal and civil, have further emphasized to me the need to present a simple guide to this topic, which is growing in importance with more use of the new technology, and more confusing and sometimes misleading if not erroneous information circulating in the popular press. This book attempts to address that need. It does not pretend to be a reference text for imaging professionals. Such books, including my own *Image Processing Handbook* (third edition, 1998, CRC Press, Boca Raton, Florida, ISBN 0-8493-2532-3) and many others, present technical references and citations to the original literature, and the mathematical underpinnings of the science. Others discuss computer programming to implement the algorithms. This book tries to show in more simple terms just what the advantages and shortcomings of digital imaging are, and how computer image processing can be used to enhance the ability to access detail in images without compromising the truth of the images.

None of the methods shown and described here is controversial, or for that matter, particularly state-of-the-art. All of them are based on well accepted and widely used algorithms and procedures, which if appropriately applied will survive *Frye* and *Daubert* challenges. It is not the purpose of this text to document the underlying science, but to explain, by example, when and why such procedures are appropriate, and what they can be expected to reveal.

The examples shown here were produced with standard and widely available computer programs. Adobe Photoshop© (Adobe Systems Inc., San Jose, CA, http://www.adobe.com) is an excellent platform for image analysis, with a consistent user interface, the ability to acquire images from many kinds of cameras, scanners and other devices, to read standard format files and print image hardcopy. Fovea Pro and The Image Processing Tool Kit (Reindeer Graphics Inc., 20 Battery Park Avenue, Suite 502, Asheville, NC 28801, http://reindeergraphics.com, ISBN 1-928808-00-X) provide the various specific algorithms for image processing and enhancement, and have been used to teach image analysis to thousands of students in workshops at North Carolina State University and elsewhere. Used together, these programs provide comprehensive tools for digital image processing and analysis.

I hope that this book will assist imaging professionals who must testify about their procedures and results to clearly and simply explain what they have done, and why, in order to assist juries in reaching a proper understanding of the evidence. For counsel, the book may suggest appropriate avenues for questioning to bring forth such explanations, and to challenge witnesses who may confuse or mislead juries (either intentionally or not) with wrong information or opinions.

John Russ
Raleigh, NC

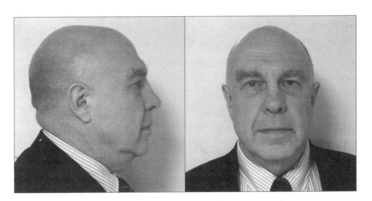

Figure 1 An unprocessed digital mug-shot of the author.

The Author

Dr. John C. Russ is an expert in image processing and analysis who has taught courses and workshops around the world, consulted for hundreds of industrial clients, and provided expert testimony in numerous civil and criminal trials.

A Caltech graduate, he was senior vice president of EDAX International (Chicago, Illinois) and Research Director of Rank Taylor Hobson (Leicester, England). Since 1979, as Professor in the Materials Science and Engineering Department at North Carolina State University, he has used image analysis and various types of microscopy in research projects, and authored 12 books and more than 300 technical articles. Short courses and workshops taught through the University's Department of Continuing Education, as well as for industrial clients, have reached nearly 5,000 students, and his *Image Processing Handbook* (CRC Press, now in its third edition) is recognized as a worldwide standard reference text for image processing and measurement. With his son, he has developed a widely used software package to implement the algorithms in the book on desktop computers. He can be reached at John_Russ@NCSU.edu.

Table of Contents

Digital Cameras and Forensic Imaging

1

Introduction

For many people, the term "digital imaging" describes the process used to paste images of Tom Hanks into old newsreels so that Forrest Gump could meet Presidents Kennedy, Johnson, and Nixon, or the even more computationally intensive efforts such as creating dinosaurs to chase human actors in Jurassic Park. These high-profile movies generate a lot of publicity for the technology used to create them, and although we willingly suspend our disbelief when we enter the theater, the knowledge that the images we are seeing do not show actual events is retained.

In fact, modern movie making uses digital imaging technology in many movies where it generates little if any publicity, and viewers may not be aware of the effects that are produced. Building movie sets and creating huge explosions is very expensive. In many productions, the sets and explosions are digitally generated scenes and the actors perform in front of blue screens so they can be superimposed afterwards, just as the TV weatherman is shown in front of a map. When real neighborhoods are used as sets, particularly for period movies, it is often necessary to remove power lines and other modern details from the images. And in many movies the stunts require guy lines and safety harnesses that are later removed from the pictures. These methods are logical extensions of techniques used in movie making long before computers, but are far more realistic than early movies in which lizards were enlarged to represent dinosaurs, or model boats in a swimming pool were photographed to produce war scenes.

Even more than movies, television advertising brings digital image manipulation into our routine daily experience. Except for the occasional

"Gosh, how did they do that?" reaction when a dancer appears to freeze in midair while the camera moves, product packages dance, or one face morphs into another (or into an animal or car), we aren't even aware of the graphic effects that are introduced. One danger in all of this exposure to digital imaging is that dramatic graphics produced by these methods become accepted as reality, so that real images of real events that lack the same level of drama and excitement don't seem as real or compelling. The second, and opposite, danger is that awareness of the ability of computer methods to produce false images that can't be easily distinguished from reality will make people distrust all images as representing truth. The widely held belief that "Anyone with a home computer and Photoshop can edit an image to introduce or exclude information" is far from true, because the skills needed are still considerable, but the process is certainly more widely available, less expensive, and harder to detect than ever before, and the contrary belief that "Seeing is believing" must also be subjected to scrutiny.

Actually, the computer graphics methods used in movies and advertising represent only a tiny corner of the more general field of digital imaging, and one that has few connections to the use of related techniques in legal proceedings. Simulations of accidents based on physical laws are sometimes used to generate imagery that demonstrates the presumed sequence of events, but these are used more in civil than criminal proceedings. Graphic rendering of the layout of buildings or rooms that use the same software used by architects to allow "walkthroughs" of proposed construction may be useful to help juries keep track of crime scenes, as an alternative to building a physical model.

The many other aspects of digital imaging include the acquisition of routine crime scene images using digital cameras rather than traditional film cameras, discussed in this chapter. Digital images have some different characteristics from film, and Chapter 2 covers the processing techniques that can be applied to them to overcome some of their limitations and produce high quality results. A rather fuzzy line separates these topics from those in Chapter 3, which use similar methods to enhance the images and extract and reveal details that may not be seen in the original. These methods are also applicable to pictures captured with traditional film cameras and later digitized to permit this sort of enhancement. This raises obvious concerns about introducing false information, so the chapter discusses in detail which procedures are appropriate and why. The closely related topic of authentication of images, including detection of digital forgeries, is also covered.

Chapter 4 discusses an application of digital imaging that has a high impact in criminal cases: the identification of individuals. Digital processing of surveillance videotapes is becoming widely used in spite of serious limitations and is discussed in terms of the factors present in recognition and identification. Other identification issues include digital aging of children's faces, and automated recognition methods using facial dimensions and iris patterns. Identification methods based on fingerprints and DNA protein separations also use images for recording, so these are discussed as well.

Chapter 5 addresses the issues confronting the technical expert acting as a witness to present information on imaging methodology and assisting attorneys and juries in understanding image evidence. This presentation is from the point of view of the technical expert, since these imaging professionals form one of the two principal groups of readers of this text. The other group is the attorneys, who must know about the technology, its possibilities and pitfalls, in order to successfully use (or challenge) the evidence and the experts. Some of these issues, including the admissibility of evidence and techniques for helping the witnesses communicate well to the jury, are mentioned within the various chapters.

Why Go Digital?

Digital cameras are taking over a major segment of the consumer photography marketplace. Only at the very high end (large format, professional cameras with interchangeable and highly adjustable lenses) and very low end (inexpensive automated snapshot cameras) are traditional film cameras holding their own. All of the major camera companies, worldwide, recognize this trend and are major players in the development and marketing of digital cameras. It is thus natural to consider the use of these cameras for, and their impact upon, the field of forensic photography. In addition, there are other sources of digitized images. Scanners now routinely capture documents for computer storage and can also be used to convert conventional photographs to digital form for processing. The signals from analog electronic devices such as video cameras used in surveillance are easily converted to digital form for computer manipulation. A new generation of video cameras that record and transmit their information digitally (DV, or digital video, format) on tiny cassettes is also available, which offer many advantages (size, image quality, ability to duplicate without loss of quality) over traditional videorecording for accident and fire scenes.

The potential advantages of digital imaging for forensic purposes are fairly obvious, if often overstated. First, the stored image can be examined immediately without any need to wait for the chemical development of the image, not even the minute or so required for Polaroid™ instant prints, so the photographer can be assured that the desired information has been captured. Second, the stored image can be transmitted via the internet, exact duplicates can be made for all interested parties, and the images can be filed archivally with no degradation. Indeed, writing images in a "tamper-proof" format such as CD-R disks is recommended to guarantee the integrity of the images. Maintaining the chain of control for evidence is thus simplified.

Finally, and most controversially, the digital image lends itself to computer manipulation to "enhance" the visibility of details. It is not always immediately clear what constitutes enhancement and what may verge into the area of improper manipulation that can distort or introduce information, and the next chapters will discuss this in detail. With the proper procedures, fully documented and well understood, it is often much easier to extract important detail from digital images than from conventional film, although many of the same results can be achieved from the latter by a knowledgeable darkroom professional given enough time.

Certainly, the procedures for image enhancement are more rapidly performed using computer-based processing on digitized images. The skills needed to run the software may appear to be easy to master, but in fact this appearance is misleading. The great danger of computer-based manipulation of digital images is that because the software seems so easy to use, and the records of the steps performed are not kept (or, if they are, can be easily lost or manipulated themselves), images may be altered in ways that introduce new information or remove original data. This may either be accomplished intentionally or accidentally, depending on the intent and skill level of the persons involved. Detecting forgeries in digital images (discussed in Chapter 3) is often much more difficult than when the same effects are produced by retouching or splicing film-based images together.

Finally, the use of the computer simply for presentation of images in the courtroom, and overlaying labels and diagrams that can be easily removed, opens up opportunities for communicating image information to juries in new ways that will challenge the courtroom infrastructure, but will be comfortable to many jurors because of their increasing familiarity with television and computer screens.

Uses of Digital Imagery

Reserving until later chapters concerns over possible misuse or misinterpretation of digital images, it is clear that the advantages of digital cameras and other image acquisition devices make them an important tool for many professional and scientific fields. Some of these are directly involved in forensic work, such as recording the crime scene and the exact location and surroundings of evidence, extracting information from surveillance videotapes, enhancing latent fingerprint and other markings on documents and other surfaces, recording imagery during autopsies, acquiring videorecordings of fire scenes for arson investigation, etc. There are also many instances in which images are used in civil trials, for example to show product failures (and consequent damage) in liability cases. Images are also used to bring into the courtroom evidence such as DNA identification by electrophoresis, close-up images of evidence obtained in the laboratory, autopsy photographs, etc., and some of the processing and presentation advantages of digital imagery discussed in later chapters may be applied to these images as well.

There are also peripheral uses of digital imaging that may never bring the images themselves into a courtroom, but are nevertheless important. For example, most microscopes (this includes light microscopes, electron microscopes, and others) used to examine particulates, tissue sections, fiber evidence, gunshot residues, etc., now record the images using digital cameras, rather than the film cameras in widespread use only a decade ago. This has the advantages of immediacy and lower cost, but also permits computer-based identification and classification of objects.

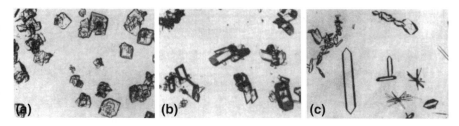

Figure 1 Identification of organic high explosives from microscope examination of particles: (a) dinitro-bis-oxamide (NENO); (b) trinitronaphthalene (TNN); (c) another form of TNN.

Such identification was performed entirely by humans until recently. For example, there are atlases of images for various types of particles (Figure 1), natural and man-made fibers (Figure 2), chemical residues from explosives, etc., that can be used to match and identify features collected in forensic investigations. (There is an enormous, but rather scattered, literature with such data. See for instance the *McCrone Particle Atlas*, McCrone Research Institute, 2820 S. Michigan Ave., Chicago, Illinois. A few other examples

Figure 2 Microscopic images of fibers: (a) cotton; (b) linen; (c) wool.

include the *Atlas of Human Hair Microscopic Characteristics*, by R. R. Ogle and M. J. Fox, CRC Press, Boca Raton, Florida; the *Forensic Insect Identification Cards*, from the Forensic Sciences Foundation, Colorado Springs, Colorado; and the *Atlas of Fibre Fracture*, J. Hearle et al., CRC Press, Boca Raton, Florida.) The procedures for accomplishing this are complex and exacting, and only a comparatively few laboratories and researchers have mastered them. Comprehensive courses in the microscopic examination of such evidence are presented regularly, for instance by the McCrone Research Institute. The use of these skilled experts is expensive and hence this type of evidence is used (by either prosecution or defense) in only some fraction of those cases in which it might potentially be helpful.

It seems reasonable to expect that the knowledge contained in these atlases and available from experts will be incorporated in computer matching programs to an ever increasing extent over the next few years. Already there are automated programs for scanning electron microscopes to search for the characteristic signature of gunshot residues on filters used to wipe suspects' hands. Similarly, there are matching programs that can identify particulates, man-made fiber cross-sections, etc. by size, shape, color, optical properties, etc. For the present, most of these methods work only in rather limited areas of application, most often in scientific research and industrial quality control where the classes of possible matches of interest are rather small (hundreds as contrasted to millions of possible candidates), but in principle they can be extended to much wider arenas, as computers get faster and the raw information is put into suitable form. Indeed, it is the effort required to assemble and encode the databases that is the principal factor delaying the more rapid implementation of these methods.

One area of successful application of automatic matching of images is the automatic fingerprint identification system (AFIS) that is now available to enable police to compare fingerprints taken from a crime scene or from a suspect with millions of stored prints, within minutes. Of course, it is necessary to digitize the image for transmission and matching. A still experimental system for palmprint identification has also been tested by several police departments. There are proposals to develop a unified national database of shell casing and bullet toolmarks to identify handguns used in crimes. Databases of paints used by all domestic and many foreign automobile makers can be used to identify paint flakes and smears from hit-and-run accidents. All of these systems use images (and for the paint flakes, supplementary compositional data taken from the scanning electron microscope) to produce very rapid identification with high confidence levels.

The algorithms used in these matching operations are complex and often not understood by the people using the instruments or computer systems. This may create difficulties for the use of such information in trial settings, as indeed similar problems have arisen in other fields such as chemical analysis when the forensic researcher may know how to use a gas chromatography-mass spectometer (GC-MS) but not be expert in the design of the equipment. Will someone who knows how to run a particle or fiber identification program be expected to understand the neural net, fuzzy logic, or principal components algorithms that it uses, and to explain why one or another of those approaches was proper in a given instance? Will bringing in another expert to explain these functions help the jury or merely prolong the trial, add to its expense, and raise the possibility of boring or confusing the jury?

Digital imaging will thus raise new opportunities for many professional fields, including forensic science, and at the same time present new challenges for the courtroom. Judges, counsel, and juries will be faced with more opportunities to be bewildered by technical jargon, bulldozed by expert witnesses, and overwhelmed by masses of exhibits. It will be largely the responsibility of counsel to determine how to use these new tools and how to challenge them.

Film as a Light Sensor

The light sensor in a film camera is a piece of photographic film. The incident light photons trigger a reaction that further chemical treatment (the development process) turns into a variation in density or color that can be viewed directly or used as a negative to produce prints for viewing. Films are available that are monochrome ("black and white") or color, and also ones that respond to infrared or ultraviolet light. Film is also used to record images from X-ray photons, used in medical and industrial imaging. The development of film materials and chemistry has been ongoing for a century and a half. The resolution of film images depends to some degree on the amount of silver halide, dyes, or other chemicals present. Usually, the greater the sensitivity to very low light levels, the poorer is the film's resolution.

Generally it is reasonable to expect the best films to have a tonal response (the ability to capture variations in light intensity) of about 4000 steps (i.e., as many as 4000 measurable steps from the darkest to the brightest recorded levels). This is usually specified as the "optical density" of the negative. Optical density is written as the negative of the logarithm of the fraction of light that is transmitted, hence an optical density of 3.6 corresponds to $10^{-3.6}$ or one part in 4000. Film typically has a spatial resolution of 3000 to 4000 points per inch (i.e., as many as 5000 discernible separate points across the width of a standard 35 mm film negative).

These values pertain to the photographic film, whether color or black and white, and whether positive or negative. Photographic prints are much poorer in both tonal and spatial resolution, and it can be difficult or impossible to show in a print all of the details that have been captured on the film. Polaroid™ "instant" film produces positive prints with even less grey scale range, as little as 10 to 15 discernible levels. This has important consequences for the use of both digital and photographic prints in a courtroom situation, which will be discussed later.

The dynamic range of film is far poorer than the human eye, which can respond to brightness changes over about nine orders of magnitude (10^9) from nearly single photon detection on a starlit night to the illumination on a sunny day on the ski slopes, but the eye cannot simultaneously distinguish all of these brightness levels. Human visual response is limited to detecting changes of brightness of about 2 to 3% (a factor of 1.02 to 1.03) that take place over a short lateral distance. More gradual changes are ignored. This works out to the typical ability to distinguish only some 20 to 30 brightness values in a scene. Human vision does not "measure" brightness but relies on immediate local comparisons to determine whether one region or object is brighter or darker than another, and the same is true for color sensitivity. The visual ability to detect a percentage change in brightness means that human vision does not respond linearly to intensity, but logarithmically, just as film does.

In terms of spatial resolution, the human eye is better than film. There are about 150 million sensors (rods and cones) in the human eye, with most of the color sensors clustered densely in the fovea (the portion of the retina which we use to "look at" things). The spatial resolution there is very high, giving the ability to distinguish marks as little as 100 μm apart. And yet we typically find that good photographs contain all of the information we can detect visually in a scene. Part of this is because our eyes don't actually communicate all of the individual sensor outputs to the brain. A very hierarchical selection and extraction process is used to find just those bits and pieces of the image that are likely to convey information, and then organize that into familiar objects. That is why people are such poor observers, particularly for unexpected objects and situations, a topic that will come up again.

The Digital Camera Sensor

While photographic film is inferior in tonal and spatial resolution to the human eye, it is quite a bit better than the detectors used in digital cameras. The devices, first used in video cameras, are much newer, and still undergoing rapid development. They are a direct outgrowth of the developments in solid-state silicon-based devices used in modern computer technology, although the most common kind (the CCD or charge coupled device) does not actually share much of the architecture of the common computer chips and requires somewhat different manufacturing technology.

The CCD chip is a two-dimensional array of diodes fabricated on a silicon wafer. Each diode is simply a light bucket. Every photon that enters the active volume of the device, which is a thin region near the surface, deposits its energy there by raising electrons from one band level to another in the atoms of the device. These electrons are trapped and can't easily get back to their lower energy state. Given enough time and some heat energy, they would, but instead the charge is transferred out of the diode and measured, thus reporting the light intensity at that location.

The detectors are arranged in a checkerboard pattern that covers the surface of the device. Typical devices in use now range from about 0.25 inches diagonal to about an inch (a little smaller than 35 mm), and on this area there may be from a half million to about 6 million individual detectors. At the upper end of this range, the spatial resolution is about one-half to one-third that of film. Sensors with up to 16 million detectors, which can match the spatial resolution of film (but not yet its tonal resolution) are being made experimentally, but appear to be a few years away from commercial availability (and will likely find their way into high cost professional cameras).

There are other types of devices that can record images. The CMOS sensor has the advantage over the CCD that it is fabricated in the same way as memory chips, which means that the cost is much lower. Also, it may in the future be possible to place some of the other circuitry (for addressing, amplifiers, voltage measurement, etc.) on the same chip for even more cost savings. But for the present these sensors are not used in high end cameras because they are much noisier than the CCD, and the noise characteristic is much more objectionable for viewing the pictures.

In the CCD, the noise is proportional to the signal (because of the fundamental statistical properties of moving the electrons from one state to another). This means that the brighter a feature is in the photograph, the more noise there is. But because vision responds logarithmically, this increased noise is not perceived as being greater and the overall result is that noise appears more or less uniform across the entire scene. In the CMOS design each detector is initially charged up like a small capacitor and then discharged by incoming light. The result is to reverse the noise characteristic so that dark areas, where vision is particularly sensitive to small changes, are especially noisy. CMOS detectors may be encountered in a few very low cost surveillance video cameras. Their primary use at this time is for videoconferencing cameras, in which a poor quality image is tolerated. However, development of these devices continues at a rapid pace and detector size and quality for both CCD and CMOS detectors are improving.

Recent misleading claims from developers that CMOS devices offer "higher resolution" than comparable CCDs are based on the fact that the sensors are smaller so that more can be packed into a given size device. This is irrelevant, since the role of optics is to focus the image onto whatever size detector is used; it is the total number of detectors in the entire image area that is related to resolution. The smaller detectors generally degrade the image because the tonal range and efficiency of the device suffers. The incorporation of on-chip circuitry to amplify signals in CMOS chips raises concerns about nonuniform response of the detectors, which leads to pattern noise in images.

One problem with having an array of CCD detectors is figuring out how to read out the charge from each of them. There simply isn't space to run individual wires to each detector, so instead they are connected to their neighbors so that the electrons in entire rows or columns can be shifted, much like a bucket brigade, to transfer the charge out to a measuring circuit (Figure 3). These devices were originally developed for use in television cameras, in which each horizontal row of detectors corresponded to one scan line in the broadcast image. In that case, the charge from the horizontal row of detectors was simply fed to an analog amplifier and became the signal. In a digital still camera design, there is still an analog amplifier followed by a high speed measuring circuit that reduces the voltage to a series of numbers that can be stored in a computer-like memory.

It is a common misconception that the number of measured values corresponds to the individual detectors. The word "pixel" takes on many different meanings in digital imagery, including the number of detectors on the chip, the number of stored values in the computer, and the number of displayed points of light on the cathode ray tube. Each of these may be quite

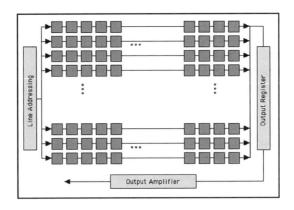

Figure 3 Diagram of a device with address logic, readout register, and amplifier.

different. The important (but not often specified) value is the number of resolution elements in the image. Enlarging the image so that this is larger than the display pixels is called "empty magnification" (the same thing that results when a photographic negative is enlarged too far). The number of detectors on the chip may be greater than the number of resolved pixels for several reasons, including the electronic characteristics of the amplifiers and the use of color filters to obtain color information from a single chip. It may also be less than the number of stored pixels, which may be advertised to imply it is the camera resolution even though this represents empty magnification. This topic will reappear several times in the following discussions.

Transferring the charge along the row of hundreds or thousands of detectors creates several problems. One is electronic noise, equivalent to the "splashing" of water in the buckets in a bucket brigade. When the chip is scanned at video rates, this noise is quite severe and sets a limit on the quality and tonal range of the resulting signal, as discussed in Chapter 4 on surveillance video. In a digital still camera the readout is much slower and there is less noise (the analogy would be to freezing the water in the buckets in the bucket brigade). In some laboratory cameras used for microscope photography, the chip is cooled to subzero temperatures to further reduce the portion of the noise due to thermal effects, but this also reduces the readout rate and is not generally practical (or necessary) for hand-held cameras.

In a typical video camera, the transfer of charge along the row of detectors occurs while light is still striking the chip, so there is also a contribution of additional light to values as they are being transferred. This is prevented in a digital still camera by the use of a physical shutter to block the light at the end of the exposure. It can also be avoided by doubling the number of detectors on the chip, covering half of them with an opaque shield, transferring the charge from the active sensors to the hidden ones, and then reading the signal out from them. This "in-line transfer" method reduces some image smearing problems in video cameras, but adds cost to the chip and reduces the light sensitivity, since half of the detector area must be made insensitive to incoming light.

The arrays are not 100% efficient in any case. Some light may reflect or scatter from the detectors, and the conversion of the photon energy to electrons is not perfect, but the main loss is due to the necessary separation between the individual diodes. The ditches that prevent charge from one detector escaping to another take up as much as 50% of the area of some of the smaller chips. Since these are usually used in consumer-grade video cameras that will only be used in bright lighting situations, the tradeoff is acceptable. For professional-grade cameras, the active detectors typically provide higher efficiency, and this can be increased by placing a small lens over each detector to collect light over practically 100% of the chip area.

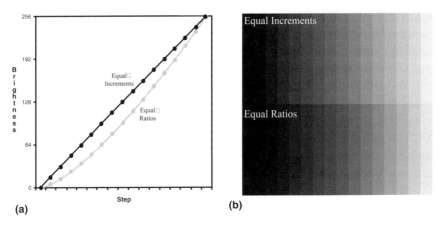

(a)   (b)

Figure 4 Brightness steps: (a) plots of intensity with equal increments and equal ratios; (b) display of the values from (a).

The output from the solid-state detectors is linear with brightness. This is an important difference from the logarithmic response of film and human vision. Figure 4 shows a series of brightness steps. One graph shows a set of linear steps (equal increments), while the second has steps with equal ratios. Note that if the smallest step on the lower graph was used to construct a linear ramp, it would take more than twice as many discrete levels to cover the entire range. This is why tonal resolution with a linear detector must be greater than for a logarithmic one, in order to show detail in the dark portions of an image. Some video cameras that use a solid-state detector incorporate a nonlinear amplifier (usually referred to as gamma, which is discussed in Chapter 2) to produce an output similar to that from tube-type video cameras, film, and human vision.

For color imaging, there are several possible strategies (Figure 5). Some video cameras use three sensor arrays, splitting the incident light with prisms so that the red, green, and blue components are recorded separately and the information combined electronically. These cameras are costly and somewhat delicate. Some digital cameras (used for microscope attachment or still photography in a studio) use a single sensor array with a rotating filter wheel, and sequentially acquire the red, green, and blue images which are subsequently combined in the computer. These are very slow, and often include a cooled chip to gain greater dynamic range; they have little or no application to hand-held or field photography.

Many video cameras and most digital cameras use a single chip with an array of detectors that have filters in front of them, so that some of the diodes receive red, some green and some blue light. Several different arrangements of colored filters are used, the most common being the striped pattern (vertical lines of red, green, and blue) and the Bayer pattern, both shown in

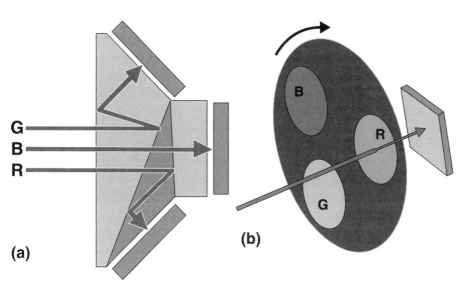

Figure 5 Schematic of three-chip (a) and filter-wheel (b) color cameras.

Figure 6. The latter has twice as many green-sensing detectors as red or blue, which gives the camera more sensitivity for green light (just as the human eye is more sensitive to green than it is to red or blue). To determine the red intensity at a location where there is no red-sensing detector, an interpolation is performed (and similarly for blue and green). Because the filters cover a broad range of wavelengths and there is some overlap in sensitivity between the various color channels, the loss of resolution is actually slightly less than would be expected. This effectively means that the camera has between half and about two thirds the resolution in each direction that the total number of detectors would suggest.

The color information may be obtained from the detectors after each line is amplified and digitized, or by using analog combinations of the signals before digitization. Some video cameras even combine the analog signals directly to produce the luminance and chrominance (brightness and color) information, using filters to smooth the signals and accomplish the interpolation of both signals from the sparse array of detectors. When a single chip camera with a Bayer pattern is used, two lines must be read out to obtain color information. This is usually done by combining lines 1 and 2, 3 and 4, 5 and 6, etc. for one interlace field, and then using lines 2 and 3, 4 and 5, 6 and 7, etc., for the other field.

A 2.1 million pixel digital camera (used in this instance to mean the number of detectors but, as noted before, in other cases this word will have different meanings) may be advertised as being able to record an 1800×1200 pixel image, but in fact that is partially empty magnification and the real resolution is less.

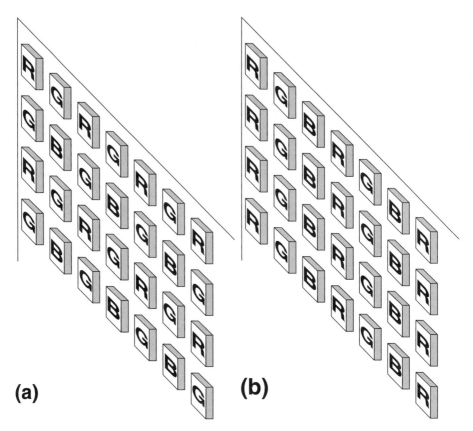

(a)

(b)

Figure 6 Diagram of Bayer (a) and stripe (b) filters for single-chip color cameras.

The tonal resolution of these devices, when read out at video rates, is rather poor. A good studio television camera (probably using a vacuum-tube detector rather than a solid-state chip, because these more expensive devices have higher resolution, logarithmic response, and a color sensitivity closer to that of human vision) should, according to broadcast video standards, be able to produce 100 distinguishable grey levels. Furthermore, these levels will be logarithmically spaced according to intensity, just as photographic film and human vision are. Solid-state detectors, however, are inherently linear devices. A solid-state video camera may produce about 60 distinguishable brightness values, but as they are linearly spaced in intensity they do not produce that many visually distinguishable brightness steps. But they don't have to, because the human watching the television image can only distinguish about 30 anyway, and the result is good enough for television, especially for camcorder videotaping (where the tape recording causes a further loss of quality). We will return to the consequences of this performance in Chapter 4, when we discuss how video signals are digitized for computer processing.

For digital still cameras, the same chip is read out more slowly and hence it can distinguish more grey levels, usually the full 256 which most cameras and computers record. Since $2^8 = 256$ it is common to refer to this as an 8 bit image, convenient to store in computer memory, which is organized into 8 bit bytes. Some digital cameras can achieve 10 or even 12 bits ($2^{10} = 1024$, $2^{12} = 4096$), but these are generally not transportable hand-held cameras, being more suitable for use in a studio setting, on a copystand, or on a microscope.

Tonal and Spatial Resolution, and their Consequences

We may take as a typical digital camera performance at this writing a chip with 2.0 to 3.5 million individual detectors, with a dynamic range of 8 bits (256 discrete grey level steps, linearly proportional to light intensity). Because these cameras acquire the color image using filters as discussed above, the actual image resolution is likely to be about 1200 × 900 pixels. (In this case, the word pixel means a resolution element in the image, corresponding to the smallest feature or dimension that can be distinguished from another; determining resolution is discussed more fully in later chapters.) A typical photographic 35 mm film negative will have a resolution of at least 3600 × 2400 pixels and, as we saw above, the human vision system has an even higher resolution capability. What are the consequences of this difference?

The primary effect of resolution is to limit the size range of objects that can be recorded and distinguished in the image. Human vision can easily distinguish objects that are millimeters in size in a scene in which other objects or distances of meters are seen. This is a ratio of about 1000:1. In a film camera, the ratio of largest to smallest objects that are satisfactorily depicted is about 100:1. This means that objects that are meters in size and ones that are centimeters in size can be recorded in the same photo, but to see the millimeter-sized details it is necessary to take close-up pictures, which no longer show the large meter-sized objects. With a digital camera the permissible size ratio shrinks again, to about 10:1 (this is a trifle conservative — under good conditions a factor of 20:1 may be achieved, but the point is the need to take more close-up pictures).

When using a digital camera, you cannot take a scene view and subsequently enlarge portions of the image so that small details become recognizable (Figure 7). The ability to endlessly enlarge or "zoom" the image magnification and extract more information is sometimes shown in movies, but does not exist in real life. There is consequently a need to take a scene view for orientation and then record many close-up pictures showing the important details, objects, marks, etc., and their placement, as shown in Figure 8.

Figure 7 Enlargement of a scene showing pixelation rather than detail.

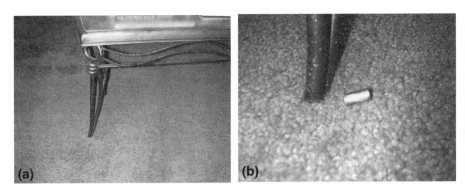

Figure 8 Example of the need for a close-up image to show detail: (a) the position of the shell casing near the table leg is documented; (b) the shell casing can be seen.

A similar restriction arises due to the more limited tonal range of the digital camera. As noted above, film can record a 12 bit image (4096 brightness levels). The digital camera can typically record only an 8 bit image (256 brightness levels), and furthermore the fact that these are linearly spaced rather than logarithmic reduces the effective tonal range by about half. A range of 100 distinguishable brightness levels is a conservative figure to use. This means that, while a single photographic film negative of a scene that is partially in shadow and partially brightly lit can be printed in the darkroom to reveal details in either the bright or dark portion of the scene, the digital image will not have as much dynamic range. In some cases it is possible to use image processing, as shown in Figure 9, to reveal the details. It will often

(a)   **(b)**

Figure 9 Example of contrast expansion to show details in a dark region: (a) original; (b) expanded contrast, as described in Chapter 2.

be necessary to change the camera exposure (aperture and/or shutter speed) to record separate images of the bright and dark regions.

Most digital cameras have automatic exposure capability and can be used in a wide variety of situations to capture visually satisfactory results. But for forensic work, it becomes more important to have fill-in flash attachments when photographing crime scenes. For events such as fires, the bright areas of flame or floodlights will tend to control the exposure so that it will be necessary to use manual settings to get satisfactory images showing the poorly lit areas (which may contain vital information). Fortunately, the digital camera preview and playback functions make it possible to determine immediately if satisfactory images have been recorded.

Taking additional pictures with different exposures, and more close-up pictures, means that in general the digital photographer will take more pictures on scene than the film photographer. That in turn means that he or she must plan ahead to have plenty of batteries (for the camera and the flash units), and plenty of storage. Most digital cameras use Compact Flash or Smart Media memory modules, while a few use floppy or hard disk storage or can be connected directly to a portable computer to store their images there. As we will see later on it is important not to use compression to try to squeeze more images into limited space. The increased number of pictures also makes it even more imperative to use a good record keeping system to identify the images as they are taken. Whether the photographer does this with pen and paper, or with a voice recorder, or in some other way, the record is vital to establish the utility of the images as evidence.

Color Response

Human color vision covers roughly the wavelength range from 400 to 700 nm, corresponding to blue through red, with the greatest sensitivity (but

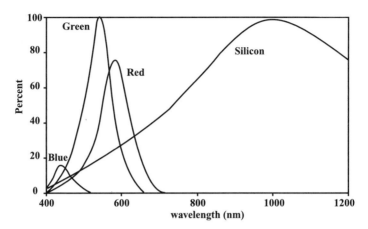

Figure 10 Wavelength responses of the human eye and a typical solid-state detector.

least ability to detect small changes in color) in the green (Figure 10). Shorter wavelengths (ultraviolet) and longer ones (infrared) are not visible. Photographic film generally covers the same range, with about the same sensitivity (of course, this is by design). Solid-state detectors do not naturally have this same coverage or response. Instead, they cover a much broader range, extending far into the infrared, and have their greatest sensitivity at the extreme red end of the spectrum, or in some cases in the infrared. Solid-state detectors are generally insensitive at the blue and ultraviolet end of the spectrum, unless special fabrication steps are used to permit light to enter from the back side of the chip. A digital camera would thus be a poor choice to record images showing the location of organic fluids revealed by "black light" (UV) illumination. The wavelength sensitivity of solid-state detectors requires some adjustment to make them useful for recording "real" color (i.e., corresponding to human vision).

Most digital cameras incorporate an infrared blocking filter in front of the chip (a few, which do not include lenses and are intended for mounting on microscopes, may omit that filter, expecting the user to install his or her own). Blocking the infrared light is important because the optics cannot focus the long wavelengths at the same point as the visible light, and images would appear out-of-focus and hazy if the filter was missing. Then, as noted above, there will be other filters used to separate red, green, and blue wavelengths for detection. These are not precise wavelengths, of course, but rather broad ranges of color that may overlap somewhat, and will vary from one camera design to another. Human vision also has three types of color sensors in the eye, which are often referred to as red, green, and blue. These have sensitivities that cover broad ranges of wavelengths and overlap considerably.

Human vision does not measure color. What one person means by yellow may be quite a bit greener than what someone else means, and furthermore each person tends to judge colors differently according to the other colors present in the scene being viewed, and according to the color of the illumination. People are quite good at comparing two colors side by side, but not at remembering colors exactly or being able to make comparisons over a period of time. Indeed, most of human vision is concerned with making comparisons rather than measurement, which will be discussed in subsequent chapters.

Color image recording, whether it is accomplished using film or digital cameras, has real difficulties in capturing the color with absolute precision. Just as with human vision, the broad range of sensitivity to the wavelengths of light that are detected by the diodes in the chip, or activate the chemical reactions in the film, means that many different combinations of colors (wavelengths of light) can produce an identical final result. On the other hand, the cameras are not sensitive to the illusions that result in human vision when adjacent contrasting colors are present. One might expect that viewing the pictures would allow a human to sense the same color information as was present at the original scene, but that doesn't quite work either, because the recorded image is much smaller than the original scene, and one component of human vision uses the surroundings and wide angle information to "correct" the interpretation of color from the central portion of the scene.

Even without considering the difficulties of interpreting the color in an image, recording an image with high fidelity color information is in itself a difficult task, especially considering the effects of incident light (which may include light reflected from other nearby, colored objects). With a film camera, the development chemistry and temperature must be carefully controlled to achieve fidelity. Digital cameras are better in that regard, but require that the red, green, and blue sensitivities be balanced for the lighting situation. This can be done (it is called making a white balance) by recording an image of a colorless (i.e., white or grey) card with the camera. Most professional and high end consumer cameras include the circuitry for performing this white balance, and tethered cameras provide software to perform the operation in the computer.

An even better method when taking pictures in which color must be known as accurately as possible is to include in the photograph a test card with known colors on it, such as a Macbeth color checker chart which has red, green, blue, yellow, magenta, cyan, and several shades of grey (Gretag-Macbeth Corp., New Windsor, New York). Using the known colors on the chart (as shown in Figure 11) with the measurements of the chart colors as recorded in the image permits correcting the colors of any other objects

Figure 11 Image of an evidence scene with a color scale included (also shown as Plate 1).

present in the scene (provided they are illuminated the same as the chart). This procedure will be illustrated in Chapter 2.

Digital Photography

Taking pictures with a digital camera is very similar to using a conventional film camera, and there is no intent to duplicate here the general background information on camera functions, which is widely available and which it is assumed the forensic photographer will already know. Exposure is controlled by a combination of aperture and shutter speed. The CCD sensor has an effective "film speed" that is typically in the ISO 50 to 200 range, about the same as most high quality films, and the exposure settings will therefore be similar to those used with a film camera. Of course, the need to use a shutter speed adequate to stop motion, or a tripod to steady the camera when using a slow shutter speed, will be the same as for a film camera. And the effect of aperture on depth of field will also be the same.

One advantage of the digital camera is the presentation of an image preview on the built-in LCD display. Most digital cameras have this facility, which shows a low resolution preview of the actual image captured by the CCD chip. This can be used to verify that the image covers the expected view, that the exposure settings are correct, and that the image is in focus. Most

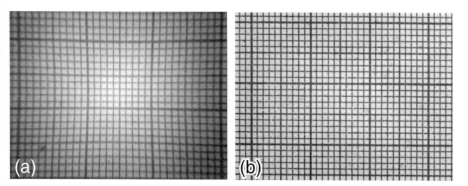

Figure 12 Example of images taken with a consumer-grade camera (a) and a professional camera with macro lens (b), showing the distortion produced by the former (also note the loss of focus and darkening at the edges).

of the professional-grade cameras include both sophisticated automatic exposure and focus circuitry, plus the capability to adjust exposure or to select aperture or shutter priority to choose an appropriate setting to achieve the desired image result. Remember of course that the goal of forensic photography is an accurate depiction of the evidence or scene and not the creation of artistic images in which defocus or motion blur may be important tools. A small aperture for high depth of field is usually preferred.

Digital cameras are currently available with prices ranging from about $100 to many thousands of dollars (Figure 12). Generally, the lower price range includes consumer models that are quite unsuitable for forensic applications (or most other professional, serious uses). The low priced models are characterized by poor, limited optics that cause image distortion and vignetting (darkening of the corners of the image), the inability to adjust exposure settings (shutter speed and aperture) and focus, only a built-in flash (support for an off-camera flash is important to enable the use of more powerful units, and to position the flash to reduce glare or sidelight surface detail for highlighting), and often even the lack of a tripod socket.

In addition, the most important single limitation that these cameras have is a small amount of memory (which is expensive), so that the images are stored in a compressed format. This is almost always done using the JPEG (Joint Photographers Expert Group) compression method, which is based on a discrete cosine transform. There are other compression schemes that are rarely used in cameras, but quite common in other computer applications, such as wavelet compression and fractal compression. Each of these techniques has somewhat different advantages for reducing the size of image files in particular cases. The goal of all compression techniques is to discard some of the image information while keeping enough cues to assist human viewers to recognize familiar objects in familiar settings. In other words, you should be able to recognize snapshots of your family and pets, be reminded of vacation sites like the Eiffel tower, and so forth. It is in fact amazing how much of the image data can be discarded — often 99% or more — while maintaining enough information for this rather limited purpose.

It is possible to achieve a modest amount of compression in an image (or any data) by finding repetitive patterns. Transmission of faxes, for example, uses a lossless compression scheme, and most computers use programs like "zipit" or "stuffit" to reduce file size for transmission. For images, this typically produces only a factor of 2 or 3 in file size. Lossy compression can achieve much higher ratios of 50:1, 100:1 or even more. But lossy compression creates a serious problem for forensic photography, because the pictures are not of familiar objects in familiar settings. It is in many cases the surface details, superficial marks, exact positioning of fragments, etc., that constitute the important evidence to be recorded in the image. This is the very information that is eliminated, modified, or shifted by the compression methods. Recording the image in a "lossless" format (often TIFF — tagged image file format — is used as a standard format recognized by many programs and computer platforms) requires much more memory for storage. The TIFF format includes provision for lossless compression based on finding repetitive patterns of pixels, but this rarely achieves more than a nominal factor of 1.5 to 2 times in file size, not the factor of 50 to 100 times possible with lossy compression. Only the higher end consumer cameras and professional cameras include or allow the addition of enough storage memory (in the form of Compact Flash, Smart Media, or a small hard disk) or provide a direct high-speed interface to the computer (e.g., a SCSI or USB connection) to allow the use of lossless storage.

A compressed image may seem to be useful to show a general view of the crime scene, to locate the close-up photographs of specific evidence. But as the photographer you would not want to put yourself in the position to answer cross examination questions like these:

Q: So you maintain that these photographs document evidence vital to this case.
A: Yes.
Q: Your camera saves these images in memory using the JPEG format, correct?
A: Yes.
Q: Isn't that what is commonly called a "lossy" format because some detail is discarded?
A: Yes, but only details that are not considered visually important.
Q: Don't you mean details that are not needed for recognition of familiar objects?

A: Yes.

Q: But these crime scene photographs are not of familiar scenes and objects. Can you show us exactly what detail has been lost in these images so that we can be assured that nothing of importance to this case was omitted?

A: No, I cannot.

If you had used fractal compression, which introduces detail into images by reconstructing them from many small copies of portions of the original, the questions would also have included, "What details have been introduced in the image reconstruction process?" and again there is no possible answer.

At this point, of course, your pictures have been eliminated as useful evidence. The simple fact is that you cannot demonstrate just what details were omitted, moved, or altered because that varies from one image to another and from one location in the image to another. In fact, a compressed image may be useful for some purposes, such as showing the general location of an object whose appearance is documented in another, uncompressed image. But by far the safest course of action is to avoid compression altogether, because it is difficult to predict when the image is acquired just what details may later prove to be important. Storing uncompressed image is not the most common procedure with digital cameras at present: many of the reports from police departments using digital cameras emphasize the "economy" of using JPEG compression to store many images on a floppy disk, or make it easy to send them via the internet. As noted above, this is not a wise choice.

Many enforcement agencies defend the current use of JPEG compression on the grounds that (a) it is widely done; (b) all recorded images have some finite resolution limit and hence introduce artefacts, and so images stored with lossy compression are still acceptable for what they do show; and (c) jurors are accustomed to looking at images and can decide for themselves if they are reliable. These are incorrect and misleading arguments.

It is certainly true that all cameras have some definite resolution limit. For film cameras the limit is usually the grain structure of the film. For digital and video cameras, whether resolution is controlled by the density of detectors in the camera, the digitization process, or the number of pixels used to store and display the image in the computer, there is a definite limit. This is discussed at length in other parts of this book. But the resolution limit imposed by the technology is uniform across the image and affects all features in the same way. Also, it is possible to calculate that resolution based on equipment specifications or to measure it from the actual image. However, compression removes features, relocates boundaries, alters the size, shape and color of features, and reduces resolution differently for different locations

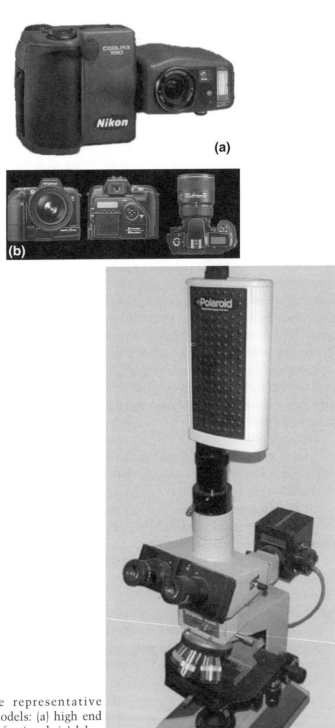

Figure 13 Three representative
digital camera models: (a) high end
consumer; (b) professional; (c) laboratory.

in the same image. In predominantly smooth, uniform regions small details are preserved better than in detail-rich areas. The JPEG method also alters pixel values differently depending on whether they lie at the center or boundaries of the 8 × 8 pixel blocks into which the image is divided. This produces artefacts and loss of detail that are not consistent and cannot be predicted. Hence it is unrealistic to expect observers to be able to judge the image quality to determine what details and features can be confidently observed. The visibility of detail in one location does not imply that similar details will be resolved elsewhere.

The argument that "everyone does it" is, of course, circular and hence meaningless. It would be more realistic to admit that everyone does it because of cost (uncompressed images require more memory to store and take longer to transmit) and because thus far no significant challenge has been raised against such images. Sooner or later images stored with lossy compression will be challenged on the basis outlined above, and the first time evidence is rejected because of compression many agencies and police departments will have to rapidly change their practice.

Some high end consumer cameras which can save loss-free TIFF images and have optics that permit adjustment of exposure and focus also include zoom lenses that have low distortion and can usually focus to less than one inch to acquire close-up macros images. These are useful for some forensic photography applications, and can serve as a useful backup camera in any case (Figure 13). Professional digital cameras are usually built on the same type of chassis as a 35 mm camera, with interchangeable lenses and comprehensive adjustability. The CCD sensor is mounted in the film plane, but is usually smaller than the size of 35 mm film. This makes the field of view smaller than the 35 mm camera, which requires some adjustment to the viewfinder and increases the effective lens focal length. This is actually a good thing, since it is the edges and corners of the optical field that have the most problems with distortion, loss of focus, and vignetting.

There are also tethered cameras that have no internal storage, but require a direct connection to the computer. These are generally intended for use in the laboratory (on a copy stand or a microscope), and will not normally be used in the field. They offer excellent resolution (both spatial and tonal) and good color stability, and allow the attachment of various lenses as needed. The digital images from these cameras may also be used in forensic situations to supplement *in situ* photos taken to document the location and surroundings of evidence which is subsequently photographed in more detail after being brought to the lab.

Maintaining Chain-of-Control with Digital Images

Digital images are simply arrays of numbers (the pixel brightness values), and can be stored with any of the storage devices used for other computer data or programs, such as tapes, disks, writable CDs, etc., provided they offer enough space for the rather large files. However, for evidence purposes it is important to use a medium that can provide security from tampering.

The requirements for image evidence can be met for traditional film by keeping control of the original negatives, preferably as an intact roll. This prevents images from being deleted, altered, or added to the set. For digital images, the equivalent security can be achieved by writing all of the images in a set to a permanent storage medium such as a writable CD-R disk, particularly one that has a serial number. It is not possible to modify, remove, or add images to this set, which can easily and confidently be copied in its entirety. Magnetic storage media, including tape and computer disks, and rewritable memory such as Compact Flash, do not have this security aspect and it would in principle be possible to edit and replace an image, remove an image from the set, etc.

CD-R and CD-RW recording is now a routine storage format for computers, and larger capacity writable DVD disks are on the horizon, but it is important to understand that these disks are not as suitable for truly archival storage as the pressed CDs used to distribute programs, movies, and audio. Those disks use pressed pits to encode the zeroes and ones that make up the information. Writable and rewritable disks use lasers to heat materials so that spots change phase or reflectivity. These disks are more easily damaged and may also be affected by heat, humidity, or solvents. Most manufacturers claim a 10 year useful life.

When image processing is performed to enhance the visibility of details in images, it is necessary to keep a careful and complete record of the step-by-step procedures used. The most straightforward way of doing this is to keep a copy of each derived version of the original image. Of course, this creates something of a numbering problem. If each image is thought of as a document, it can be assigned an identifying number (usually called a Bates number — the sequential seven-digit number that is conventionally stamped onto paper documents and indexed). Of course, it is the index that enables one to make sense of all this, and because the numbers will be chronological in sequence, a separate record of some kind will be needed to track the changes from one image to the next.

For some cases, and for some attorneys, this offers an important fringe benefit. One strategy of dealing with discovery proceedings is to flood the opposing side with a huge barrage of documents, in which are buried the few that might actually matter. The expectation is that opposing counsel will not be able to discover from the flood which strategies are considered important and which documents provide the key and relevant facts. With image processing, it is quite simple to apply every conceivable operation to every recorded image, in various sequences, to produce a host of images most of which will never be used again, but which hide in their profundity the few important ones that will be relied upon to prove a case. (This doesn't feel entirely ethical to a scientifically trained mind, but it does seem to appeal to the adversarial mind-set of some attorneys I've worked with.)

In any case, it is vital that as the imaging expert you maintain your own record (which will be protected from discovery as your work product) of the images and processing steps employed to obtain the results about which you will testify. I find it helpful to use a graphic aid for this, since there are typically many threads that branch and merge to lead from some original images to some final conclusions, but each person will evolve a methodology that fits best with their own style. Don't expect to be able to keep the information in your head.

Digital Video

The important aspect of digital photography (as emphasized above) is not the solid-state detector, but the storage of the image information in a digital form that permits duplication, transmission, storage without degradation, and facilitates processing. Some of the same advantages can be achieved for video recording by using digital video (DV) technology. Conventional analog videotape recording writes the signal to magnetic tape in much the same way as audio cassette tape is recorded. The saturation of the magnetic field at each location on the tape is proportional to the voltage of the signal, and writing, storing, and reading the tape introduces small but cumulative distortions in the information.

The most commonly used video format (VHS) mixes the brightness (luminance) and color (chrominance) information in the same signal, which requires reducing the bandwidth and resolution of the color. This is considered an acceptable tradeoff for video, based on the fact that human vision is not generally confused by colors that bleed across edges, or are not perfectly reproduced, particularly when following motion or recognizing familiar objects in familiar scenes. As will be discussed in Chapter 4 on surveillance video, the actual resolution of VHS analog video is quite poor.

The light detector used in most portable video cameras and camcorders is the same CCD array as that used in digital cameras. The readout rate is much faster, to achieve the 30 frames per second rate used in video (25 frames per second in the European video standards), and this introduces more noise to reduce the tonal resolution of the data. The interlace method that scans even numbered video lines in one sixtieth of a second and then goes back and fills in the odd numbered video lines in the next sixtieth of a second can introduce artefacts that limit the vertical spatial resolution, and can cause moving objects to be distorted. But it is the analog recording that represents the greatest limitation to quality.

The newest development in video cameras and camcorders is DV (digital video) which records the video signal in digital form. A useful comparison to audio recording is that CD (compact disk) audio technology records the signal digitally. Audio CD recordings do not degrade with time, and the disks can become scratched or dirty without causing loss of information. The DV signal is still recorded magnetically on oxide-coated tape, but instead of the magnetic saturation being proportional to the voltage, the voltage has been converted to a series of numbers and the binary representation of the numbers is recorded as a series of regions where the field is oriented in one direction or the other. Since the magnitude of the saturation is not important, the digital recording is much more stable and robust. These recordings, and the transfer of the same set of information to computer memory, can provide the same advantages as digital photography.

The density of recording is very high on DV tapes, which are tiny cassettes. The camcorders that use this technology are small and easily transported (Figure 14). Some can fit into a pocket, while the slightly larger ones accept interchangeable lenses. All have sophisticated exposure (aperture and shutter speed) controls, live displays using lower resolution LCDs that can also playback the recordings, and built-in computer interfaces that can transfer the digital data at high speed (IEEE 1394, sometimes called Firewire, which is built into some modern computers). The color and brightness information are not intermixed, as they are in VHS (but not in S-VHS) analog recordings. Also, DV cameras use a progressive scan rather than the usual video interlaced scan (alternate 1/60th second fields for the even and odd scan lines) which improves image quality, particularly when recording motion. The cameras can usually be connected to conventional video displays, in which case the internal progressive scan data is converted to interlace, and the resolution is lowered to match that of conventional video.

The downside to DV recording is that the data are compressed. The most basic compression step is to reduce the sampling of color information relative to brightness (analog video does this as well, based on the relative insensitivity of human vision to color boundaries). Then each video frame is compressed using a JPEG-like algorithm. Some implementations also use an MPEG (Moving Pictures Expert Group) compression scheme which looks for redundancies from one frame to the next. Any lossy compression method can alter

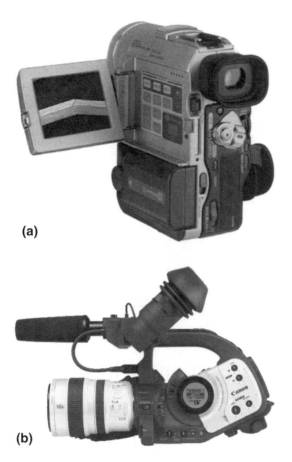

(a)

(b)

Figure 14 Two representative DV camcorders: (a) pocket type; (b) professional.

fine details of the image, so all of the same cautions as noted above for the compression of digital still images apply.

However, even with the lossy compression, the image resolution of DV video is still much better than that of VHS analog video. VHS recordings can, under ideal conditions, achieve 240 resolution elements across the screen width for a color image. DV more than doubles that, to about 500 resolution elements. Note that this is still much less than the number of detectors on the chip, which for a typical DV camcorder is more than 750 (but if a single chip detector is used, color filters are employed to separately detect red, green, and blue light in different locations, which must then be interpolated). High end DV cameras use a compression scheme that does not reduce the color sampling as much, and have three chips with a light prism to separate the R, G, B images. They also generally have better optics and/or accept inter-changeable lenses, have comprehensive exposure adjustments, etc.

Figure 15 One frame from a video recording of hurricane damage.

A DV camcorder would be a better choice than a VHS camcorder for recording crime and fire scenes in which motion is important, and may be a useful accessory for recording the overall appearance of a crime scene to show the layout and locations of detail images by panning. This is also true for fire or accident scenes, and for surveys of property damage (Figure 15) video recording is an excellent tool. The usual strictures for any video recording apply, of course. The photographer should avoid rapid pans and zooms, which are disorienting. It is wise to turn off the sound recording (unless separate microphones are being used to pick up important sounds from the scene) to avoid recording comments and other noises from the photographer which are apt to be distracting and perhaps misleading (but would be duplicated along with the video evidence if they were captured on the tape).

It would not, however, be acceptable to use a DV camcorder to record close-up details in lieu of acquiring still photographs with a digital or film camera, because of the poorer resolution and concerns about compression.

Scanners

The flat bed scanner has become a widely used computer accessory for acquiring digitized images from photographs and other flat objects. The most frequently encountered forensic laboratory applications for scanners are dig-itizing X-ray films (both medical and industrial) and one- or two-dimensional gels as used in electrophoresis and other protein separation technologies. They may also be used in some situations to convert conventional photographic film to digital form, for example to facilitate transmission, enhancement, or measurement.

Flat bed scanners come in a variety of types. Some have the light source on the same side of the object being scanned as the detector and can be used only for reflected light imaging of opaque objects such as photographic prints. Others, which generally cost more, have a second light source on the opposite side of the object and can record transmitted light, thus being able to scan negatives. It is always preferable to scan the original negatives rather than prints, because the negatives have a much greater dynamic range (tonal resolution) as well as superior spatial resolution, and, of course, because they are the more original source of evidence. Special scanners are available specifically for scanning photographic film such as 35 mm slides and APS rolls.

The light detector in a scanner is a single linear array of CCD detectors, rather than the two-dimensional array used in camera chips. This linear array may cover up to the full width of a page with 600 or more sensors per inch (dedicated film scanners use much finer sensor arrays). It is scanned slowly along the document imaging one line at a time, with the spacing of the lines also about 600 per inch to yield an overall resolution of 600 dpi (dots per inch — another phrase which, like "pixels," has more than one meaning in different situations). Some scanners have higher spatial resolution, up to more than 1000 dpi, but many advertise an "interpolated" resolution that is much greater than the spacing of the actual sensors. This empty magnification is not real resolution and should not be used to judge the quality or performance of the device. Scanners usually have high speed digital interfaces to the computer (SCSI or USB are the most common) and produce very large files containing all of the individual pixel values.

For color imaging, some scanners use three sets of detectors to capture red, green, and blue light, while others use a single detector array and insert filters or turn on different light sources as the array is scanned to capture the different colors. Some scanners do this during a single pass over the object, while others make three passes, once for each color. Each method has both advantages and drawbacks. The three pass method is slower and places demands on the mechanical precision of the scanner to align the three separate images. But the single pass method with different filters or lights gives less stable illumination and makes the color fidelity poorer, and the use of three sets of detectors means that they have different view angles onto the sample, which produces color offsets and shadows, particularly for objects with rough surfaces or with a finite thickness, such as color film.

Some inexpensive consumer-oriented scanners have only an 8 bit (2^8 = 256 levels) dynamic range, but the slower readout from these devices as compared to a digital camera (a scan can take a minute or more) usually allows images to be captured with greater dynamic range. For color images, each channel of red, green, and blue intensity is recorded with the same dynamic range. Twelve bits (2^{12} = 4096) per channel is common, and is required to deal with the dynamic range of X-ray films in particular. The specification may also be given as the maximum optical density that can be measured, and 1 part in 4000 corresponds to an optical density of 3.6. Some X-ray films contain a large amount of silver and so can capture larger density differences, as for example between bone and tissue in medical X-rays or between metal and cracks or gaps in industrial X-rays. Some dedicated medical X-ray scanners provide 14 bits (2^{14} = 16384 shades of grey, or an optical density of 4.2).

The detector output is linear, but many of the applications of scanners are to photographic materials whose response is logarithmic, so the transfer of the data linearly into computer memory is appropriate and introduces no additional complications for interpretation.

There is another way to obtain digital representations of photographs taken with film. Kodak offers a service through its film development centers that uses their scanners to digitize photographic negatives and deliver them along with (or in lieu of) traditional prints. Kodak Photo CD would seem to be an attractive, inexpensive way to obtain these files, which can be read into many computer image processing programs. It would be logical to expect the professional scanners to be of the highest quality and well maintained. However, the Kodak Photo CD format uses a lossy compression algorithm to pack more images onto the CD, and so for the same reasons discussed above this would not be a preferred method for images in which details are important. It has been suggested in some texts that Photo CD recording of conventional film photographs provides a low-cost method to acquire digital images, but for the same reasons as presented above for avoiding compression, it is wise to avoid this.

Taking Good Pictures

There is no intent to reproduce here a comprehensive course on how to become a skilled photographer. It is assumed that the reader already has the basic skills but is new to digital photography, or that he or she will refer to texts such as the following to learn about the particular requirements of recording images of crime scenes, fires, evidence, etc.:

D. R. Redsicker, 1994, *The Practical Methodology of Forensic Photography*, CRC Press, Boca Raton, FL, ISBN 0-8493-9519-4

S. Staggs, 1997, *Crime Scene and Evidence Photographer's Guide*, Staggs Publishing, Temecula, CA, ISBN 0-9661970-0-3

L. S. Miller, 1998, *Police Photography*, 4th edition, Anderson Publishing, Cincinnati, OH, ISBN 0-87084-816-X

This has much more to do with accurate recording of the information about each picture, and acquiring images that reproduce the details of the scene, evidence, surface markings, etc., than it does with the usual concern of photography, which is the taking of attractive pictures.

It is a common misconception, which attorneys often try to urge upon juries when images are presented that are damaging to their case, that "everyone knows how to see things in pictures" so that no explanations, and preferably no close-ups, processing, or enhancement are needed. This is not true for a variety of reasons. First, people are not particularly good observers and only a very small amount of the information in the raw images they normally see is retained or interpreted. Second, human vision is comparative in nature: the brightness, color, size, distance, etc., of several objects or persons can be judged rather well when they are seen side by side with the same lighting and without distracting surroundings. But many common optical illusions demonstrate that this comparative capability can be easily tricked, and that comparing one object or person to a remembered image is fraught with peril.

Photographs can serve simply as a reminder of events and things seen. But they can also bring to our attention details of a scene that were not noticed and perhaps not visible originally. It is the role of the photographer to accomplish both of these purposes: to make the crime scene visible to the jury, and to highlight important evidence which may not have been readily observed in that scene at the time.

There are a few points that are worth including here, because of the particular needs of digital cameras and the possibilities of digital image processing. First, remember that the digital camera has less resolution (both spatial resolution and tonal resolution) than a typical film camera. This means that you will have to take more pictures. Separate pictures are needed that capture different parts of the scene with an appropriate exposure setting so that regions are not too bright or too dark to show detail. With film, it is often possible to depend on darkroom techniques to bring out the detail in the dense or thin regions of the negative, but digital imaging provides less latitude for under- or over-exposure.

Also, more separate pictures are needed at different scales. Overall scene images will show the placement of objects, but close-ups will be required to show their exact relationship to their immediate surroundings, and to reveal surface details and markings. With film, it is often possible to enlarge a wide-field negative to reveal this fine scale information, although better results may be obtained by adjusting the lighting, using indirect or bounce flash to eliminate bright reflections, or to use grazing light to highlight surface marks. Digital imaging provides less latitude for enlargement and (as noted above) places more restrictions on illumination and exposure.

With any photographic recording of evidence, it is wise to include in the scene, placed close to objects of interest, a reference scale. Normally this would be, at a minimum, a ruler with clearly legible markings, to indicate size. If color is of potential importance in the evaluation of evidence, it is also wise to place a reference color scale close to the objects of interest, so that it has the same illumination that they do.

When the size, shape, or spacing of objects or marks (e.g., footprints or tire marks, as shown in Figure 16) is important, it is important to try to capture images that have a normal or perpendicular view of the interesting portion of the scene. For instance, a photograph taken straight down onto footprints, with an appropriate scale in the image and low angle lighting that makes the marks evident, can be used to measure the size of the shoe. It is sometimes difficult to acquire such ideal images, with either a film camera or a digital one. Large objects or ones that are placed high or far away must sometimes be photographed as best they can, even if there is some foreshortening or perspective distortion involved.

As will be shown in the next chapter, it is often possible to rectify images and remove perspective distortion so that measurements can be made, but it requires that an appropriate target scale be included in the image. More than a simple ruler is needed. The target must be placed so that it will have the same foreshortening and perspective distortion as the subject of the photograph. An ideal situation would be to surround the object with a clearly visible square, perhaps formed by placing crime scene tape on the ground.

Figure 16 Images of a footprint taken with perpendicular (a) and oblique (b) viewing angles.

Knowing the dimensions of the square (which of course should be written down in the photographer's log book) makes it possible to process the image to obtain the same view that a normal or perpendicular image would have shown.

When a complete surrounding outline is not practical, a specialized target may be included in the scene that can be measured by the computer and adjusted to its true shape, with the remainder of the image receiving the same adjustment. This works fine for planar (flat) surfaces, but will distort objects that stick up from that surface. Also, the method requires that the lenses used must not have significant barrel or pincushion distortion, and sufficient depth of field to keep all of the image in sharp focus.

In some cases involving irregular surfaces, it may be worthwhile to take stereo pictures for later measurement. This simply requires taking two photographs with the camera position shifted sideways by a known distance, as shown in Figure 17. For "true" stereo viewing, the distance should be about 7 cm (the average distance between human eyes), but for measurement purposes a larger distance (which must be measured accurately) can be used.

Presenting Pictures in the Courtroom

Photographs are often introduced into evidence in the form of 8 × 10 inch prints, which can be conveniently handled, and overlaid with transparent sheets that can be marked to indicate features of interest. It is important to remember that even for the case of photographic film, the quality of the print is considerably inferior to the original negative (which should properly be introduced into evidence as well, in case comparison is needed). The process of making the print from the negative involves selections of paper, chemicals, processing, and exposure that can emphasize some of the information in the image while de-emphasizing other information, another reason that the negative should be available.

Figure 17 Example of a stereo pair image.

For digital images the making of hard copy prints is more controversial and troublesome. There are computer printers that can produce results with much of the quality of a photographic print. These include dye sublimation printers and some of the newest ultra-fine drop ink jet printers, when used with special coated paper that prevents the ink from penetrating the fibers. These devices produce quite good quality prints for viewing and have the same life (estimated at tens of years) as photographic prints, but have less tonal range than photographic prints; in addition, the cost of materials and the time required to produce multiple copies are not less than with traditional darkroom methods. There are also a few specialized printers that use thermal activation of laminated dye layers, and even devices that print onto silver halide media (the same as traditional photographic materials). Most of these devices use quite expensive consumable materials, for a cost-per-print of several dollars, and many can only produce small print sizes.

But the most widely available, faster and cheaper alternatives, such as "regular" ink jet printers or even laser printers, will be used in many cases, with very inferior results. Producing enough copies for distribution to all interested parties also encourages the use of these faster printers or even color copiers. Conventional ink jet printers do a good job of recording computer graphics, which generally have bright, saturated colors and which usually lack fine detail. If the prints are made on plain paper, the colors and sharpness suffer. Most ink jet prints fade rapidly with temperature or exposure to light. Laser printers approximate the halftone printing used in newspapers and magazines to record grey scale images (color laser printers are not quite as good as color halftone printing, and also are chiefly useful for bright graphics rather than photographs). Color copiers are markedly inferior in color fidelity and saturation to the originals.

It is amazing what people will put up with in terms of the desire to have hard copy output of images they have seen on the computer screen. The poor quality of many of these prints is tolerated because people are pleasantly surprised that they can get any hard copy at all, and the print serves primarily as a reminder of what they saw on the screen. The resolution of these prints is generally inferior to the screen display. Even a photographic print produced using a film recorder and a darkroom, which is extremely expensive, does not have the dynamic range of the display (although the negative from which the print is made may capture the full tonal range of the display).

Most computer printing technologies do not deposit grey scale or color information on the paper that corresponds to each display pixel on the computer screen (dye sublimation printers and a few ink jet printers do). Instead, many ink jet printers and all laser printers use a set of fine dots deposited from three or more different colored inks to build up the color of each display pixel in a small area on the paper. The ink dots are very small

Figure 18 Enlarged view of printed halftone color image (see Plate 2).

indeed: many printers have more than 600 and sometimes more than 1000 dots per inch. But it requires a considerable array of these dots to simulate a range of brightness and color values. For each color, an array of 10 × 10 dots, some of which were omitted, could be visually combined to produce 100 different degrees of brightness. Most computer screens can display 256 degrees of brightness each for red, green, and blue at each point.

Furthermore, in most cases the color dots cannot be in the same place, but are instead carefully placed next to each other (Figure 18). A few printers (for instance, dye sublimation printers) do blend the dots to generate mixed colors, requiring special inks and/or paper. But in all cases, the printing deposits ink that darkens the white underlying paper, and it is the paper that contributes the brightness to the image. So you can't have a fully bright, fully colored spot on the hard copy, as you can on the computer screen. The hard copy printout is always darker, duller, and with much less dynamic range than the screen image.

This makes it quite difficult to generate acceptable printouts from digital images after they have been transferred to the computer. The common alternatives are to photograph the images from the computer screen and then print them in the traditional way (which obviates many of the original advantages of digital imaging in the first place), or to use expensive printing devices such as film recorders, dye sublimation printers, or ink jet printers with special papers and inks, to generate hard copy.

The apparently simple approach of setting up a camera on a tripod facing the computer monitor and taking photographs of the screen, which are subsequently printed, should be avoided. Unlike a film recorder, this produces several kinds of distortions which can be used to invalidate the images as evidence. Most computer monitors have curvature in one or both directions, and in addition placing the camera far enough away to avoid fisheye lens distortion and exactly in line with the monitor's axis is very difficult, so the images will be spatially distorted. In addition, unless you are prepared to answer questions about the monitor's phosphors, what the gamma setting for the brightness scale was and how it was recorded by the film, how the monitor color settings were calibrated, and so on, this approach is very likely to raise admissibility problems. Film recorders use calibrated color wheels to expose each frame separately with red, green, and blue signals to produce traceable, distortion-free images. However, they are expensive and slow devices, and the problems of variation in making prints from the negatives remain.

There is another possible solution, but it will require some changes in courtroom procedures and infrastructure. Many people now feel quite comfortable in viewing information directly on television and computer screens. It should be possible to set up monitors where all concerned can view them and show the images directly from computer memory. All of the same advantages of being able to superimpose marks to indicate regions of interest remain (give the witness a mouse to draw on the screen, just like the annotation of sports events on TV). The obvious problem is how to preserve this testimony and evidence in the court record. There are certainly ways that the images and the marks can be recorded (e.g., on a writable CD), but this is new ground for trial records and will no doubt be tested as to its acceptability.

Another, somewhat less satisfactory, method is to use a projector. Many courtrooms allow 35 mm slide projectors to be used to present photographs, which is good because these images coming directly from the film are superior in many ways to photographic prints. It seems a small step to replace the slide projector with a computer projector, which takes the same signals that are fed to the CRT display and uses them to drive small LCD or micromirror arrays to project the same image onto a screen. These images can be quite bright with nearly as much contrast and saturation as the CRT exhibits, but of course much larger. It requires some control over room lighting for best viewing, and of course does not answer the question of recording the testimony and images.

Summary

The use of digital photography in place of conventional film photography imposes some new technology and new requirements on the recording of forensic images. It also opens up some new possibilities and conveys some advantages. The advantages have to do primarily with the digital recording of the data, which permits duplication, transmission, and storage of the images without any degradation of quality. The possibilities have to do with processing and enhancement, discussed in the following chapters.

The new technology is based on arrays of solid-state detectors with appropriate circuitry to measure the charge deposited in each detector by the incident light. The new requirements reflect the fact that these detectors have less resolution than film. This is true both for spatial resolution — the number of points in the image that can be distinguished from each other — and tonal resolution — the number of brightness values that can be distinguished.

Ancillary to digital still cameras, flat bed scanners and digital video cameras also have a role in forensic image recording. Digital images also present a challenge to the normal use of hard copy photographic prints in trial proceedings.

For the photographer the usual skills of recording meaningful and true representations of crime and fire scenes, autopsies and other medical images, objects (both *in situ* and in the lab), surface markings, and other evidence remain important. In addition, it is necessary to understand the possibilities and limitations of the tool so that the proper information is available from the digital images, scales and color normalization charts are included in the images, and adequate notes are kept for documentation.

Processing Digital Images

2

One of the advantages of using digital images (either acquired using a digital camera or by converting an analog video signal or photograph to digital form) is the possibility of using computer-based image processing. This carries with it concerns over removing or, worse, creating detail in the images that may alter their effect as evidence. Certainly, there are procedures that can be used that may cause such alterations. In this chapter and the next, the methods to be reviewed are those which do not cause such problems. They can be conveniently subdivided into three groups:

1. Methods that deal with and attempt to correct the specific defects present in digital images, resulting from the hardware used to acquire them, discussed in this chapter.
2. Techniques that enhance the visibility of image detail in ways analogous to darkroom processing of conventional photographs, discussed primarily in this chapter.
3. Enhancement techniques that are not easily performed with film or photographs, but which can be carried out using computer-based methods on digital images, discussed in the next chapter.

Chapter 3 will also deal with detection of inappropriate processing, characterization of image quality, and detection of forgeries and other improper image manipulation.

Noise in Digital Images

As discussed in Chapter 1, digital cameras sample the incoming light using a solid-state chip. The photons are captured to produce an electronic signal that is subsequently measured. The processes by which the photons produce electrons, the electrons are moved about, and the electronic signal is amplified, are each accompanied by the introduction of noise that is superimposed onto the signal. The "signal-to-noise ratio" is a measure of how much noise has been mixed into the final image.

It is probably important at this point to qualify the term noise. It takes on different meanings in different situations, generally being used to describe any part of the total signal that does not contain the "useful" image information that we desire. It is convenient mathematically to deal with a particular type of noise called Gaussian random noise, and indeed many of the processes that take place in the chip and in the electronics are of this type. The term random means that each pixel in the image is affected independent of every other pixel, and Gaussian means that the noisy contribution to the signal can be either positive or negative (brightening or darkening each pixel) with a distribution that corresponds to the familiar bell curve or normal curve encountered in statistics. The normal or Gaussian curve is important in statistics because it results whenever there is a large number of independent sources of variability which combine together to affect the final result (e.g., a plot of the height of individuals in a population).

To see this random Gaussian noise, one can acquire a digital image of a uniform grey scene (a standard grey card used by photographers can be used, but simpler expedients will also work). Figure 1 shows such an image. It is accompanied by the image histogram, a very important tool that will be used extensively in this chapter. The histogram is a plot of the number of pixels in the stored image that have each of the possible values of brightness, typically from 0 to 255 (for an 8 bit image). Ideally, the uniform grey image should have all of the pixels with the same exact value, corresponding to the scene brightness. In actuality, the histogram will show a peak that has a finite width, and a shape similar to the Gaussian or normal curve.

The width of the peak is a measure of the noise content of the image. As shown in the figure, images acquired under different conditions may contain differing amounts of noise. Generally, for a digital camera, images taken in dim light which require more amplification of the signal will show more noise than those acquired in bright illumination. Also, for color images, the blue channel will typically be noisier (less uniform) than the red or green channels, because the sensitivity of the solid-state chip in the camera is lower for blue light and, hence, more amplification is required. In these cases, it is a combination of the physics of producing electrons from photons and the signal amplification that produces the greatest contribution to the noise.

In other circumstances, there are other dominant noise sources. Video cameras produce images that are noisier than digital still cameras because of the need to rapidly transfer the signal from the individual diodes in the chip (the bucket brigade described in Chapter 1). It is common to find that the

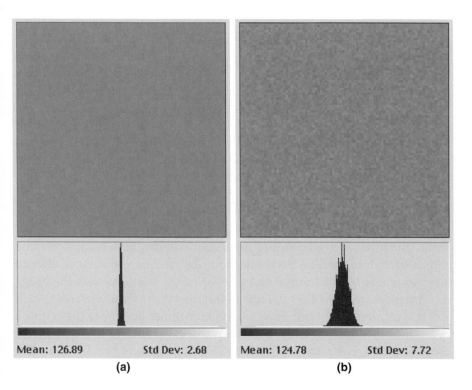

Mean: 126.89   Std Dev: 2.68   Mean: 124.78   Std Dev: 7.72
                 (a)   (b)

Figure 1 Two images of a uniform grey target (an expanded portion to show individual pixels), with the corresponding image histograms: (a) low noise, which cannot be easily visually discerned in the image; (b) greater noise, which produces a visible speckle. The latter image was acquired with low illumination and the brightness increased as discussed later in this chapter.

noise is greater on one side of the image than on the other, because the electrons were passed through more intermediate diodes before reaching the output amplifier. When digitizing video signals, random or not-so-random variations in the high speed analog-to-digital converter (the "framegrabber" or ADC), or digital noise picked up from the computer circuitry, can also add to the total signal noise.

It is not always possible to identify or quantify all of the individual sources of random noise, but the effect is usually to produce a significant variation in the values for stored pixels that represent features in the original image that were the same in brightness, and this variation is often Gaussian in shape or close to it. This phenomenon is not unique to digital photography. Film also has variations in density within uniform regions, based on the variation in the size and number of the silver halide particles or dye molecules in the film. The number of such particles or molecules is rather large and the relative variation is rather small, and in any case it is not usual to measure the film's density, so the noise is not noticed or considered to be a problem.

In many cases with digital cameras and adequate scene illumination, random noise variation in the signal is not an important problem for digital images either. The small variation in pixel brightness represented by the breadth of the histogram peak may not be visually evident in the images and can be ignored. But particularly for video images (discussed in Chapter 3) and for the dark portions of scenes that are not uniformly lit (e.g., shadow areas or dark areas in fire scenes), the presence of random noise can result in a visible "speckle" in the image that is objectionable.

There are several available methods for dealing with this noise. In some cases (usually only of practical interest in a laboratory setting) it is possible to increase the illumination, or add together many images of the same static scene (the random noise tends to cancel out, improving the signal-to-noise ratio), or reduce camera electronic noise by cooling the camera chip to subzero temperatures. In most cases, however, we must deal with the image as it has been acquired and attempt to improve the noise afterwards.

It is very important that noise reduction methods be applied to the image before any of the enhancement methods discussed later in this chapter are employed. Otherwise, the result of the enhancement will be to increase the visibility of the noise. This is visually distracting and makes it much more difficult to see the real features and particularly the important details in the image. If the noise in the image is so great that it cannot be satisfactorily reduced, then no subsequent enhancement of the image to reveal the detail will be possible.

Noise Reduction Methods

Averaging together a series of images of a static scene (or, depending on the design of the camera, using a longer exposure time to capture more signal) improves the signal-to-noise ratio because the signal continues to add from all of the individual frames while the random noise variations tend to cancel out. If temporal imaging is not possible, then spatial averaging is often employed to average each pixel with its neighbors. In uniform regions of an image, this produces a similar effect. Figure 2 shows the result, using the noisy image from Figure 1 with a filter that averages the 21 pixels in a circular neighborhood 5 pixels across. Like all neighborhood filters, this approach uses the original pixel values in the stored image to calculate a new array of pixel values, applying the same logic to every individual pixel.

Special rules can be employed to deal with the borders of the image, where the number of neighbors is reduced. None of the various methods is

Mean: 124.59   Std Dev: 1.85

Figure 2 Neighborhood averaging on the image of Figure 1b, showing the reduction in pixel variation.

uniformly satisfactory, and the best procedure is to avoid trying to use pixels that lie within the radius of the neighborhood used from the picture borders. Typical neighborhood sizes range from 3 × 3 up to about 15 × 15. There are few situations in which larger neighborhoods are useful, and when they are it is usually more efficient to perform the procedure using frequency space or Fourier space transforms as discussed in Chapter 3.

The problem with neighborhood averaging as a noise reduction technique is that it only works inside uniform areas. At a line or edge of some feature in the image, the assumption that all of the pixels in the neighborhood represent samples of the same brightness level that can be averaged together is clearly false. The result is to blur edges and narrow lines, which makes them more difficult to detect visually and also shifts them, altering the dimensions and shape of features in the image. Figure 3 shows an example of this blurring effect.

Better results can be achieved by calculating a weighted average of the pixels in the neighborhood. Instead of just adding the pixels together and dividing by the number, a set of fractional values that add up to 1.0 can be used. Multiplying the numerical weights times the pixel values and adding them up to calculate the new pixel value is called a "convolution." The best set of weights (for the purpose of reducing random noise while minimizing edge blur) has the shape of the bell curve. This is properly called Gaussian

Figure 3 A noisy image: (a) original, acquired by available light at dusk with the brightness increased; (b) smoothed by neighbor averaging using a 5 pixel wide circular neighborhood; (c) smoothed with a Gaussian filter, standard deviation = 0.8 pixels; (d) smoothed with a median filter using a 5 pixel wide circular neighborhood.

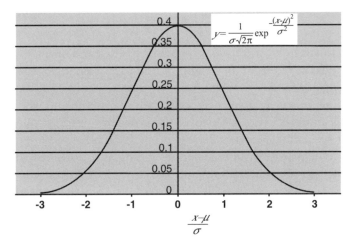

$$y = \frac{1}{\sigma\sqrt{2\pi}} \exp^{\frac{-(x-\mu)^2}{\sigma^2}}$$

$$\frac{x-\mu}{\sigma}$$

Figure 4 The Gaussian distribution or bell curve, showing the standard deviation.

weighting, and is characterized by the standard deviation of the curve. As shown in Figure 4, the standard deviation is the radial distance at which the curve drops to 68% of the central or peak value. The overall radius of the neighborhood is about three times the standard deviation. Table 1 shows some typical filters.

Historically, computers had more difficulty in performing real number arithmetic than in using integers. For this reason, many systems implement weighted averaging (and the other neighborhood convolution operations described below and in Chapter 3) using integer values. This requires that the set of neighborhood weights, often called a "kernel," be written as a set of integers to be multiplied times the pixel values. The sum of the products is then divided by the sum of the weight values, producing an approximation to the actual Gaussian values that is usually acceptable. Table 2 shows integer filters that correspond approximately to a Gaussian.

The Gaussian filter is often implemented in digital image processing systems as a two-pass operation that blurs the image in the vertical direction and then in the horizontal direction. This is possible because the Gaussian filter is "separable" as the product of two orthogonal, linear operations. Most neighborhood convolution operations cannot be speeded up in this way, and for purposes of understanding the methods used it is more general and convenient to deal with the full array of weight values.

In Figure 3, the effect of using a Gaussian filter on the same noisy image subjected to simple averaging shows less blurring for an equivalent amount of noise reduction. It can be shown mathematically that the Gaussian filter provides the optimum results using convolution with a set of neighborhood weights for removing random noise from a noisy image. However, there is still some blurring of edges and loss of image sharpness which we would prefer to avoid.

Median Filtering

A fundamentally different method of dealing with pixels within a neighborhood is based on rank ordering. Instead of multiplying all of the pixel values in the neighborhood times weight values, rank ordering lists the pixel values in order of brightness. The noise reduction method based on this ranking then selects the median value from the list. This is the central value in the list, which has the same number of brighter and darker neighbors. This median value becomes the new value for the pixel at the center of the neighborhood.

Median filtering is a superior noise reduction method, because it does not blur or shift edges (as shown in Figure 3). Since multiplying and adding is a very fast computer operation, while ranking is not, median filtering can be slower than convolution. However, with modern computers the difference is trivial and median filters are widely used for noise reduction. This method is also able to remove some other kinds of noise, as discussed below.

The adjustable parameter for the median filter is the size of the neighborhood used. As the radius is increased, the degree of noise reduction is increased, as shown in Figure 5. However, a large neighborhood will eliminate

Table 1　Gaussian Smoothing Filters with Standard Deviations of 0.5, 1.0, 2.0, and 4.0 Pixels

a) sigma = 0.5 pixels

```
0.3182  0.1171  0.0058
0.1171  0.0431  0.0021
0.0058  0.0021  0.0001
```

b) sigma = 1.0 pixels

```
0.0796  0.0620  0.0293  0.0084  0.0015  0.0002
0.0620  0.0483  0.0228  0.0065  0.0011  0.0001
0.0293  0.0228  0.0108  0.0031  0.0005  0.0001
0.0084  0.0065  0.0031  0.0009  0.0002
0.0015  0.0011  0.0005  0.0002
0.0002  0.0001  0.0001
```

c) sigma = 2.0 pixels

```
0.0199  0.0187  0.0155  0.0113  0.0073  0.0042  0.0021  0.0009  0.0004  0.0001
0.0187  0.0176  0.0146  0.0106  0.0069  0.0039  0.0020  0.0009  0.0003  0.0001
0.0155  0.0146  0.0121  0.0088  0.0057  0.0032  0.0016  0.0007  0.0003  0.0001
0.0113  0.0106  0.0088  0.0065  0.0042  0.0024  0.0012  0.0005  0.0002  0.0001
0.0073  0.0069  0.0057  0.0042  0.0027  0.0015  0.0008  0.0003  0.0001
0.0042  0.0039  0.0032  0.0024  0.0015  0.0009  0.0004  0.0002  0.0001
0.0021  0.0020  0.0016  0.0012  0.0008  0.0004  0.0002  0.0001
0.0009  0.0009  0.0007  0.0005  0.0003  0.0002  0.0001
0.0004  0.0003  0.0003  0.0002  0.0001
0.0001  0.0001  0.0001  0.0001
```

d) sigma = 4.0 pixels

0.0053	0.0052	0.0049	0.0046	0.0041	0.0036	0.0030	0.0024	0.0019	0.0015	0.0011	0.0008	0.0006
0.0052	0.0051	0.0049	0.0045	0.0040	0.0035	0.0029	0.0024	0.0019	0.0015	0.0011	0.0008	0.0005
0.0049	0.0049	0.0046	0.0043	0.0038	0.0033	0.0028	0.0023	0.0018	0.0014	0.0010	0.0007	0.0005
0.0046	0.0045	0.0043	0.0040	0.0036	0.0031	0.0026	0.0021	0.0017	0.0013	0.0010	0.0007	0.0005
0.0041	0.0040	0.0038	0.0036	0.0032	0.0028	0.0024	0.0019	0.0015	0.0012	0.0009	0.0006	0.0004
0.0036	0.0035	0.0033	0.0031	0.0028	0.0024	0.0020	0.0017	0.0013	0.0010	0.0007	0.0005	0.0004
0.0030	0.0029	0.0028	0.0026	0.0023	0.0020	0.0017	0.0014	0.0011	0.0008	0.0006	0.0005	0.0003
0.0024	0.0024	0.0023	0.0021	0.0019	0.0017	0.0014	0.0011	0.0009	0.0007	0.0005	0.0004	0.0003
0.0019	0.0019	0.0018	0.0017	0.0015	0.0013	0.0011	0.0009	0.0007	0.0005	0.0004	0.0003	0.0002
0.0015	0.0015	0.0014	0.0013	0.0012	0.0011	0.0009	0.0007	0.0005	0.0004	0.0003	0.0002	0.0002
0.0011	0.0011	0.0010	0.0010	0.0009	0.0008	0.0007	0.0005	0.0004	0.0003	0.0002	0.0002	0.0001
0.0008	0.0008	0.0007	0.0007	0.0006	0.0006	0.0005	0.0004	0.0003	0.0002	0.0002	0.0001	0.0001
0.0006	0.0005	0.0005	0.0005	0.0004	0.0004	0.0003	0.0003	0.0002	0.0002	0.0001	0.0001	0.0001

Note: In all cases only the lower right corner of the symmetrical weight array is shown. The first value in the table corresponds to the weight for the central pixel in the neighborhood.

Table 2　Integer Filter Weights that Approximate Gaussian Smoothing with a Standard Deviation of 1 Pixel

0	0	1	1	1	1	1	1	1	0	0
0	1	1	1	2	2	2	1	1	1	0
1	1	2	2	3	3	3	2	2	1	1
1	1	2	3	4	4	4	3	2	1	1
1	2	3	4	4	5	4	4	3	2	1
1	2	3	4	5	5	5	4	3	2	1
1	2	3	4	4	5	4	4	3	2	1
1	1	2	3	4	4	4	3	2	1	1
1	1	2	2	3	3	3	2	2	1	1
0	1	1	1	2	2	2	1	1	1	0
0	0	1	1	1	1	1	1	1	0	0

Note: The sum of the products must be divided by the sum of the weights = 233; usually this number is kept below 256 to facilitate rapid arithmetic, and whenever possible is made equal to a power of 2. The entire filter array is shown.

Figure 5 Application of a median filter to the image from Figure 3: (a) 3 pixel wide neighborhood; (b) 11 pixel wide neighborhood; (c) 21 pixel wide neighborhood; (d) 3 pixel wide neighborhood, repeated 10 times.

real details in the image (points or lines) which are smaller than half its size. It is necessary to use a neighborhood size that is small enough to retain real details, and to repeat the operation several times in order to reduce the noise. If the details are as small as single pixels, then by definition they cannot be distinguished from the noise and cannot be removed. This is, of course, part of the reason that digital images are limited in their ability to show small details or to be enlarged as much as film-based photographs.

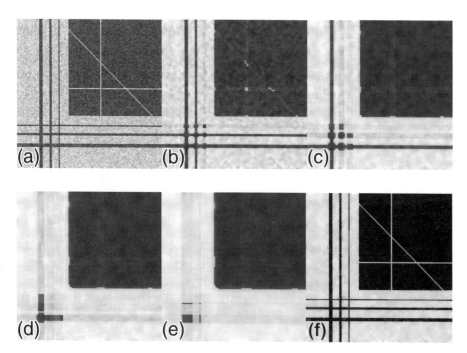

Figure 6 Test image containing fine lines and noise (enlarged to show pixel detail): (a) original; (b) conventional 3 pixel wide median filter; (c) conventional 5 pixel wide median filter; (d) conventional 7 pixel wide median filter; (e) conventional 9 pixel wide median filter; (f) hybrid median filter (see text).

Many implementations of the median filter use a square neighborhood rather than a round one, in order to simplify the programming required. This can cause slight variations in the image and distortion of some edges and corners, but it is usually not a major problem. One difficulty that occurs with both round and square neighborhoods is that corners of features are slightly rounded, as shown in Figure 6. In the figure, narrow lines are removed by a median when the neighborhood becomes large enough that the light or dark values from the line can never contribute the median value. If the lines are real image detail, they should not be removed. But if the lines are scratches or other noise defects, their removal is appropriate and correct.

A modification of the technique, called a hybrid median, is capable of preserving narrow lines and corners. In the hybrid median, several sets of pixels are ranked separately, and the median values are then ranked amongst themselves to select the final value. In the example shown in the figure, the central pixel and the four neighbors forming an X are ranked, the pixels forming a + are ranked, and the median values from those two rankings and the original central pixel are ranked, with the median value being assigned to the pixel. For larger neighborhood sizes, more orientations can be included.

Processing Color Images

Neighborhood filtering methods based on either convolution with weights or ranking must be applied with care to color images. As discussed in Chapter 3, if sharpening operations (also called high pass filters) are applied separately to the red, green, and blue channels of the image, the new pixel values will generally vary in proportion in the derived image. The result is to introduce colors into the final image that are quite different from those present in the original, which is visually distracting and potentially confuses the evidence represented by the picture. The proper way to apply sharpening techniques to color images is to first convert them from the RGB space to a space that separates the intensity from the color information. This is discussed in Chapter 3.

For smoothing operations, it is generally acceptable to work with the RGB planes. However, it is important to know that the noise characteristics of the three color channels are generally different. In most cases, the lower sensitivity of the detector to blue light (and for a single chip camera, the fewer detectors assigned to capture blue light) results in more noise in that channel. This is considered acceptable for the design of cameras because human vision is also less sensitive to blue light, but it requires some additional care in processing. Examining the histogram of a uniform grey card captured with a color camera typically shows results like those in Figure 7. The breadth of the peak indicates that there is much more noise present in the blue channel, and slightly more present in the red channel, than in the green channel.

For noise reduction purposes, this suggests that the green channel should be smoothed least and the others correspondingly more. In the image shown in Figure 8, this has been done. The image was acquired using no flash in dim light, and then brightened as discussed below. This produces visible noise in the pixels. A Gaussian smooth with a standard deviation of 0.25 pixels was applied to the green channel, a Gaussian smooth with a standard deviation of 0.5 pixels to the red channel, and (because of the significantly greater noise) a median filter using a 7 pixel wide circular neighborhood was applied to the blue channel.

It is also possible to perform a median filter operation directly in color space. This requires some redefinition of the concept of a median, as discussed in Chapter 3. The pixel that is "most like" its neighbors is taken as the median and the color values assigned to the central pixel in the neighborhood in the derived image. Figure 8c shows an example.

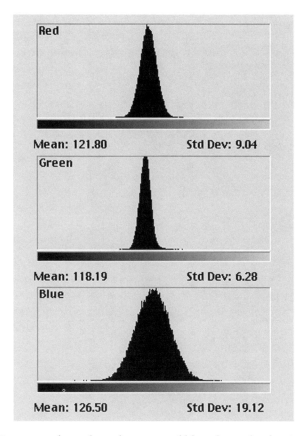

Figure 7 Histograms from the red, green, and blue channels of an image captured from a grey test card in dim light.

Nonrandom Noise

Random Gaussian noise is used as a model for noise in imaging systems because it is mathematically straightforward, and because when there are many independent sources of noise, their combination tends toward the Gaussian case (mathematically, this is known as the "central limit theorem"). When it is present in images it generally appears as a characteristic "speckle" pattern in which pixel brightness values vary slightly. But in many situations there is one dominant source of noise that is not Gaussian in nature and produces different effects in images, which can be more visually distracting than Gaussian random noise.

"Salt-and-pepper" noise is characterized by black and/or white pixels scattered at random throughout the image. It can result from "dead" pixels in the detector chip, from electronic noise during the transmission of signals,

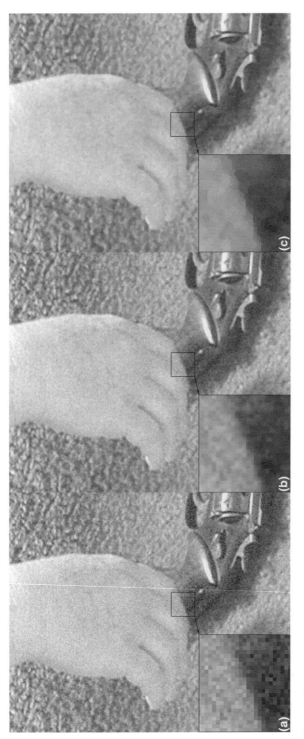

Figure 8 Fragment of a color image enlarged to show pixel noise: (a) original; (b) noise reduced by processing individual RGB planes as described in the text; (c) noise reduced with color median as described in the text. (Also shown as color plates 3, 4, and 5.)

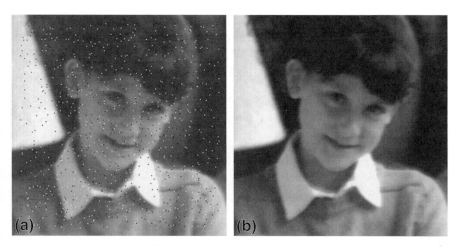

Figure 9 Removal of salt-and-pepper noise: (a) image with random black and white pixels; (b) application of a median filter (3 pixel wide neighborhood).

or from spots on recording tape where the magnetic oxide has been removed through wear or other damage. Usually the spots are single pixels, which can be effectively eliminated with a median filter, as shown in Figure 9. Essentially, this selects one of the surrounding pixels which is most likely to have the missing value and substitutes that for the black or white speck.

Pattern noise (sometimes called fixed pattern noise) results from variations in the detectors in the chip. Particularly with long exposures of several seconds, which are sometimes needed in dim light surveillance situations, fire and accident scene imaging, etc., the differences between individual detectors can produce bright or dark spots in the stored pixel array that are reproducible for a particular camera, exposure time, and temperature (which affects the characteristics of the detectors). As shown in Figure 10a, this usually shows up as colored noise pixels since different detectors are used to capture the red, green, and blue light. When this type of noise is encountered, it is sometimes feasible to remove it by capturing a "blank" image (covering the lens and taking a second exposure with the same exposure settings). This will have the same pattern noise as the original, and can be used to remove the noise by subtracting the blank or background image from the scene image, pixel by pixel and color channel by color channel, with results as shown in Figure 10b. In this particular case, a color median filter would also have eliminated the pattern noise.

Line noise is represented by narrow light or dark lines, often horizontal or vertical, in the image. It can result from scratches on movie film, retrace or synchronization problems in video images, or a defect in the readout electronics in the chip. In this case, a median filter with a neighborhood that is larger than the width of the line can be used, and will replace the pixels with others from nearby. However, the ideal neighborhood would be one that was restricted in

Figure 10 Removal of pattern noise due to detector variations by subtracting a blank image: (a) original image (enlarged), showing colored pixels resulting from 8 second exposure time; (b) result after subtracting a blank image collected for the same time (also shown as Plates 6 and 7).

its extent along the direction of the line. Some systems allow creation of custom neighborhoods that can be used to eliminate line noise (Figure 11), and some have them built in for the most common cases of vertical or horizontal lines. Scratches or cracks in a document, or video line dropout in an image, can be "repaired" quite effectively with such a filter (Figure 12).

Video signals, particularly ones from inexpensive cameras, or recorded on consumer-quality VHS recorders, or digitized with inexpensive framegrabbers (analog-to-digital converters) are susceptible to line jitter or interlace offset.

Figure 11 Constructing a neighborhood for median filtering that will remove scratches oriented at 45 degrees from upper left to lower right.

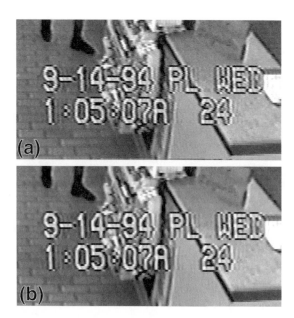

Figure 12 Removal of video line noise using a median filter with a vertical neighborhood: (a) image with horizontal line noise; (b) application of the median filter.

This can occur due to internal electronic or mechanical defects, or can result from motion in the image which displaces objects in the odd- and even-numbered scan lines. This causes the horizontal scan lines in the image to be offset in their starting points and, hence, displaced sideways in the image, as shown in Figure 13. Usually it is possible after acquisition of the digital image to shift the individual scan lines to line the features in the image up properly, but doing this manually by eye is very time consuming, especially when fractional pixel shifts are needed. Automatic methods using cross-correlation are possible but rather specialized. This method is usually performed using the Fourier-space representation of the image, which is discussed in Chapter 3.

Fourier space is also used for removal of high frequency components of periodic noise. This can arise from electronic interference, including poor grounding of the video camera that can introduce beat frequencies from the power supply. In some cases periodic noise can also result from vibration. It results in the superimposition on the signal of high frequency sinusoidal variations that appear on the image as one or more sets of lines that march across the image. These are closely related to the moiré patterns that are present when printed (halftone) images are scanned into the computer. The procedures for removing these patterns are shown in Chapter 3.

Adjusting Contrast, Brightness, and Gamma

The contrast range from the brightest to darkest parts of a scene is often greater than can be shown in photographic prints, even though it may be

Figure 13 Video image of a moving vehicle, showing interlace offset.

captured on the negative. This problem is also present in digital imaging, but because of the greater dynamic range of film the problem of capturing the important image detail so that it can be viewed is worse for digital images. It is important whenever possible to use control of the lighting to obtain the best possible initial image. In Figure 14 a scene showing the exterior and interior of a car shows little or no detail in the shadow areas in front of the passenger seat. If a flash had been used at the time the image was recorded (Figure 15), the presence of a gun in that area would have been evident.

Figure 14 Image of a car in bright sunlight, with the interior in shadow.

Figure 15 The same view as Figure 14, but using fill flash to illuminate the interior.

Obtaining optimum original images is always preferred, but sometimes this does not happen. The image from Figure 14 can in fact be processed to make the details in the shadow area visible. One way to do this would simply be to expand the contrast for the dark area (Figure 16) by setting the white and dark limits to clip most of the bright areas of the image to white. This is equivalent to printing the photographic negative with overexposure of the bright areas, but of course the process is faster and somewhat easier to control using the computer. The process alters the contrast and brightness of the image, but it is usually simpler to think of it as simply setting the black and white points for the display of the image. Areas that are brighter than the white point are shown as white, with all detail lost, and areas that are darker than the black point are shown as black, with all detail lost.

Even better results can be obtained by adjusting the gamma value of the displayed image (Figure 17), by adjusting the position of the middle grey point of the image. This creates a nonlinear relationship between the recorded

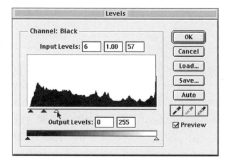

Figure 16 Adjusting the black and white limits on the histogram of Figure 14, and the resulting expanded contrast.

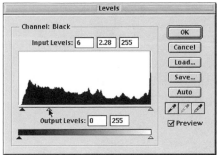

Figure 17 Adjusting the gamma value by shifting the mid-grey point on the histogram of Figure 14, and the resulting altered contrast.

brightness and the displayed brightness, and is equivalent to making a photographic print on paper of a different hardness grade. This is called a gamma adjustment because the same term has historically been used for the slope of the density vs. exposure curve for photographic materials; the density is proportional to the logarithm of the incident light intensity.

The actual numerical values are generally of no importance. Just as in the photographic printing of a negative, it is primarily a visual process to determine the best settings to show details in the image without eliminating any of the other details present. Of course, it is important to keep a separate copy of the unmodified image, which can be examined to assure that nothing of importance has been accidentally lost, for instance by clipping to white or black.

Note, however, that while the image shows the presence of a gun in the area in front of the passenger seat, there is insufficient image resolution to enlarge the image to show the details of the weapon for identification (Figure 18). Proper procedure would also require a close-up image of the weapon. (The original picture was obtained with a 2 megapixel digital camera.)

Adjusting Color Balance

Color values are recorded in a digital image as intensities of red, green, and blue as captured by the sensors and converted to stored numbers, typically in the range from 0 (black) to 255 (maximum). The values are a complex function of the spectral characteristics of the incident light on the scene, the color characteristics of the features present (generally the way that light absorption and reflectivity vary with the wavelength of color of the light), and the sensitivity of the color sensors themselves. The first and last of these are very difficult to characterize or standardize. Light sources ranging from incandescent or fluorescent indoor lighting to natural sunlight have very

Figure 18 Enlarged portion of Figure 17 (right) showing the limited resolution of detail due to the size of individual pixels.

different distributions of intensity vs. color. Some lighting, such as from low pressure sodium street lamps, may be essentially monochromatic. Even in laboratory situations, the lighting on a microscope or copy stand varies significantly with minor changes in the voltage applied to the lamp or with the age of the filament.

Likewise, the filters used to select wavelengths of light that reach the R, G, and B sensitive detectors in the camera do not have sharp cutoffs. Instead, they pass rather broad ranges of colors with attenuation that drops off gradually at the edges of each range. This means that the same RGB values may be obtained with different colors present in the scene and different illumination or, conversely, that the same colors in the scene may produce different measured RGB values under different lighting conditions. Similarly, if color film is used to record images which are later scanned into the computer, the color values depend on the film sensitivity and the illumination. Using a color film balanced for sunlight with incandescent lamp illumination, for instance, produces images with a pronounced color shift (and vice versa).

For human vision purposes these color shifts make surprisingly little difference, because we compare different colors within the same scene, but do not "measure" the color or do a very good job of comparing colors in a scene to ones that are remembered from a previous one. Varying illumination is compensated for by assuming that certain familiar objects have the expected color, or by using information from the periphery of our vision. In addition, human judgment of colors is strongly affected by the presence of other colors nearby in the same scene. Many visual illusions rely on these effects and demonstrate the difficulty people have in accurately judging color.

Figure 19 Image of a steer standing in a field, showing the effect of rotating the hues by about 25 degrees (Plates 8 and 9).

Digital cameras are not subject to illusions, but it is still necessary to make some adjustment for the incident illumination. This is usually referred to as performing a "white balance," and some cameras (especially video cameras) allow you to capture an image from a neutral grey card and specify that as having no color. This allows the camera to compensate to some degree for the illumination, but does not compensate for light from one part of the scene falling onto another object, or for color shift effects that can occur in very bright or dark areas.

The exact color information in crime scene photographs is not usually of great importance. The important evidence can be collected and the purpose of the images is to identify the objects and record their locations. If comparisons between objects are needed, they can be performed later in a laboratory setting and the colors compared under the same lighting. In some cases, it is important to be able to compensate for lighting variations between two images. The simplest approach, which is sometimes adequate, is to perform the adjustment in a hue-saturation-intensity space (described in detail in Chapter 3). The intensity (brightness) and saturation (amount of color) are not altered. The hue, representing the angle in the color wheel as discussed later in this chapter, is shifted to rotate the colors, for instance moving the yellows toward either green or red, as shown in Figure 19.

A second approach is a direct analog to the method used in film photography, which is the insertion of compensating color filters in front of the light source when making prints from film that has been exposed with a color imbalance. The vector mathematics of this adjustment is slightly more

Figure 20 Image of cars in a parking lot: (a) as photographed; (b) after adjustment of R, G, and B intensities to make the paving neutral grey (Plates 10 and 11).

complicated than the simple hue rotation and is more easily understood as proportional changes in all of the red, green, and blue intensities in the image. The best way to apply this method is to find some feature in the image that should be colorless (white or grey) and adjust the RGB weighting to remove any color present.

In the example of Figure 20, the original image has a color cast that makes the true colors of the cars difficult to determine. If we can assume that the paving is really a neutral grey with no color, then adjusting the intensities of the red, green, and blue to achieve this balance also restores true colors (or at least more nearly correct colors) to the rest of the image. After adjustment it is possible to determine that the pickup truck is green, the SUV is black, the car is red, and the station wagon is blue.

These two methods do not work well in all cases. The first one fails for colors lying in other parts of the color wheel (different hues), because the effect of incident illumination on color imaging is more complex than a simple uniform shift applied to all hue values. The hue variation is not constant, and the saturation and brightness may also be affected. The second

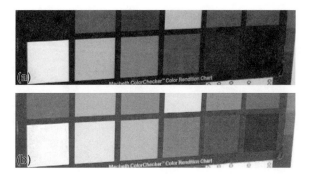

Figure 21 A portion of a color chart: (a) original image; (b) after tristimulus correction using the values in Table 3c. (Also shown as Plates 12 and 13.)

approach does not correctly adjust both bright and dark colors, since the recorded intensity values of red, green, and blue are not linear.

A more general approach is called a tristimulus correction. It requires recording a color standard under the same illumination conditions as the original image. The example shown in Figure 21 is a digital image of a portion of a Macbeth color checker chart. Three of the color swatches are known to be red, green, and blue. In the particular lighting conditions present when the image was recorded, the measured intensities show significant values of red, green, and blue in all three areas. In other words, the interaction of the incident light and the color patches has produced a complicated spectrum of scattered and reflected light, from which the color filtering used in the camera has allowed some light to reach all three detectors.

Knowing that the measured intensities should be pure red, green, and blue allows the calculation of a 9 × 9 matrix of values called the tristimulus color corrections. The procedure is to first form a matrix of the measured intensities in each of the RGB channels for each of the three color patches, as shown in Table 3. Then this matrix is inverted. When the inverse matrix is multiplied by the measured RGB values it produces maximum intensity for R, G, and B in each corresponding patch, with no contribution from the other colors. Multiplying the pixel values throughout the image by the tristimulus correction matrix produces an image in which the colors are corrected. Notice in the example that the three other color patches (cyan, magenta, and yellow) and the grey scale steps are also adjusted when this is done. Notice, too, that some of the corrections are quite large in this example. Generally, it is wise to use strong incident illumination when acquiring images for this approach, in order to avoid the imprecision that results from very small (dark) intensity values.

In all cases in which color adjustment of an image is used, it is important that it not be restricted to only one portion of an image or one object within it. Making such adjustments to an entire scene is a correct response to the

Table 3   Calculation of Tristimulus Color Correction

a) Measured Average RGB Intensities in Red, Green, and Blue
Target Areas of Original Image

	Measured Intensity	Red	Green	Blue
Area:	Red	90.223	45.865	35.556
	Green	59.933	86.941	40.296
	Blue	38.088	36.635	43.028

b) Normalized Intensity Matrix (Values from **a** Divided by 255)

0.353826	0.179863	0.139435
0.235031	0.340945	0.158024
0.149365	0.143667	0.168737

c) Inverse of Matrix **b**

4.94557	−1.46508	−2.71469
−2.27990	5.52036	−3.28586
−2.43661	−3.40327	11.12704

need to deal with nonstandard lighting or camera response. Human vision is quite tolerant of lighting variations and makes its own "white balance" compensation quite well in most cases. But altering the color information in part of a scene or in one object alters the relative color information in ways that produce false information for the viewer, who compares the color information between objects in the same scene.

Adjusting Size (Magnification)

When comparing images of presumably matching objects, such as shoes and footprints or tires and tire treads, it is generally necessary to adjust the size of the images to be the same. When done in the photographic darkroom this is non-controversial and just involves adjusting the size of the prints so that the scales recorded in each image are the same. Similar steps can be performed using digital images and computer software.

As an example of the kinds of processing that can be performed on digital images that are direct analogs of those carried out in the darkroom for conventional film negatives, consider the following example. An image has been acquired of a footprint (Figure 22). The usual care has been taken to photograph it with a perpendicular view to eliminate distortion, and the original image includes a scale, to establish dimensions. Because the scale is white, and there are some bright stones in the picture, the overall range of tonal values is rather large and the contrast for the details in the footprint is not great. Enhancement of contrast (Figure 23) can be performed by setting

Figure 22 Image of a footprint, with scale.

the bright and dark limits of the image to just encompass the range of tonal values in interest in the footprint. It would also be permissible to adjust the gamma of the image if this improved the visibility. These actions are equivalent to selecting a printing paper of the desired hardness grade and adjusting the exposure and development time to maximize the image contrast in a traditional darkroom print, as discussed above.

Based on the tread pattern, it is determined that the boot is a particular brand, and that information is used as part of the search for the perpetrator. When a suspect is arrested, a search turns up a pair of boots with the same tread pattern. A picture is taken in the laboratory of the tread on the left boot (Figure 24). Note that the scale in this case has been raised to the same level as the tread, so that it accurately represents the dimensions of the tread pattern. However, the image is not taken at the same magnification as the original footprint and so the relative size of the two images must be adjusted using the two scales. On the original footprint, measurement of the scale gives a calibration value of 93.25 pixels per inch. On the picture of the boot sole, measurement of the scale gives a calibration value of 79.67 pixels per inch. Consequently, in order to facilitate comparison, the boot sole picture is reduced to 85.4% of its original size (79.67/93.25 = 0.854). Also (Figure 25) the contrast of the image was increased in the same way as for the footprint, since only the sole of the shoe is actually important.

Reduction of size is usually a better procedure than expansion because the latter requires interpolation of values between actually recorded pixels, with the possibility that details of features in the image whose size is not changed will not have matching information in the enlarged image (the interpolation cannot insert details that were not recorded to begin with). Also, in this case, the picture of the boot sole must be flipped and rotated so

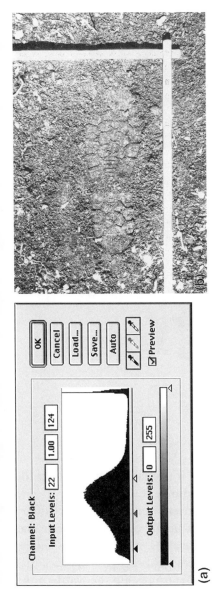

Figure 23 Contrast enhancement of Figure 22: (a) setting limits on histogram; (b) result.

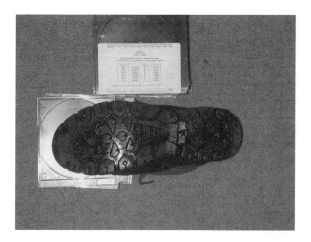

Figure 24 Image of a boot sole, with scale.

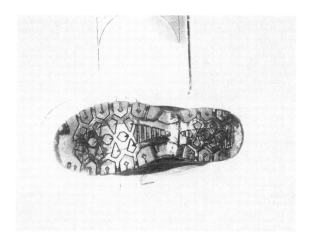

Figure 25 Contrast expansion applied to Figure 24.

that it will match the footprint. At this point it is easily confirmed that the two images show the same sole design. But do they really match?

With a digital imaging system it is easy to overlay one image on the other. This is equivalent to superimposing two film negatives on a light box and allows the same facility for positioning one relative to the other. As shown in Figure 26, this operation reveals that the length of the boot matches that of the footprint (the shoe is the same size as the print), but the width of the boot is considerably greater (the shoe has a greater width). This is exculpatory evidence: the boot does not match the footprint and presumably the boot's owner is not responsible. If two photographic prints were held side by side, or only the length of the boot was compared to the footprint, the impression

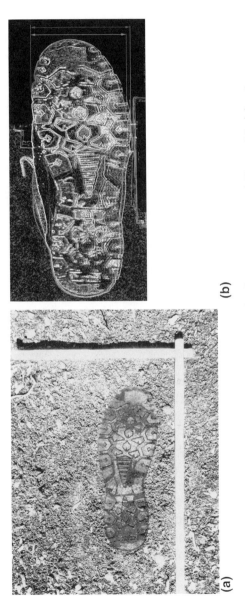

(b)

(a)

Figure 26 Superposition of the boot sole from Figure 25 onto the footprint image of Figure 23: (a) using transparency; (b) using color channels (shown at Plate 14).

Figure 27 Inch scales from Figures 22 and 24, after adjusting the magnification as described in the text.

would be that they matched and this important evidence would be missed. Placing the two images, adjusted to the same scale and aligned, into different color channels also produces a useful direct visual comparison of the shoe widths.

When different image magnifications must be adjusted to match, it is often helpful to show the adjusted scale markers as well, to confirm that they are the same (Figure 27).

Spatial Distortions

Recording images in the correct orientation is important and strongly preferred in forensic applications, because it greatly simplifies subsequent comparisons and analysis. Sometimes it is physically impractical or for one reason or another cannot be done. If a suitable scale is established around the object or feature of interest, digital image processing can be used to correct the image afterwards and obtain the desired view. This type of operation would be very difficult to accomplish with traditional film, requiring setting up the film in the original camera orientation and printing onto an inclined paper (assuming that enough depth of field in the optics could be obtained). In practice, it is very rarely done.

In the example shown in Figure 28, a footprint has been photographed at an oblique angle. This would be impossible to compare directly to an image of the boot, and in fact would be difficult to measure to determine size and width, even with a standard scale in the picture, because the magnification is different in different locations within the image, and measured in different directions in the image. However, in this example a scale has been placed in a rectangle around the print. Notice that the rectangle on the ground is represented as an irregular four-sided polygon in the image (no two sides are the same length nor parallel to each other).

This type of foreshortening can be corrected using simple projective geometry, provided that the surface is known to be flat. In the example shown in Figure 29 a perspective correction has been applied by marking locations at the corners of the marked polygon which are known to

Figure 28 Footprint image taken at an angle.

Figure 29 Perspective correction applied to Figure 28.

correspond to the four corners of a rectangle, and then using a computer calculation to stretch and compress portions of the image. The result can now be measured and compared correctly. For larger regions, outlining the region to be imaged with crime-scene tape can be used. This process is sometimes called orthorectification.

The limitation of this method is that any feature extending out of the plane of the polygon will be distorted in the perspective adjustment. For example, in Figure 30 the front of a building has been imaged with the expected perspective distortion. When this is adjusted (Figure 31) the dimensions of the windows and floors in the building are corrected, but the lamppost, which actually sticks out over the street perpendicular to the face of the building, is distorted and appears to lie at an angle. Also, the trees appear to

Figure 30 Image of the front of a building.

Figure 31 Perspective correction applied to Figure 30.

be in front of the wrong windows, because they do not lie in the plane of the front of the building. This limitation is inherent in the method.

Other Applications

Images taken from known positions or incorporating scale in the field of view can also be used for photogrammetric measurements. Based on simple trigonometry, measurements of distances in the projected image of the photograph can be converted to true distances in the scene. If many such distances are determined, then they can be used to construct a simple three-dimensional (3D) recreation of the scene (Figure 32). More elaborate software that generates "virtual" views from a database of 3D coordinates and covers the surfaces of the objects created with the actual images recorded by a digital camera can be used to produce "walkthrough" scenarios in which the jury or other observers are able to view a crime or accident scene from

Figure 32 Simple reconstruction of a crime scene by combining digitized images (courtesy Dr. Samuel Rod, Bristlecone Corp.): (a) to (d) individual photographs; (e) composite.

various points of view. The software used is designed for architectural rendering purposes. Constructing the necessary databases is tedious and expensive work, used in only a few high profile cases for the novelty value of using the interactive computer graphics.

Another application of spatial distortion of digital images is the "aging" of pictures of missing persons, especially children. The National Center for Missing and Exploited Children has applied this technology to many missing children cases with excellent results. Also, the Louisiana State University Forensic Anthropology and Computer Enhancement Services Laboratory (FACES) has performed these operations in various law enforcement cases when children and adults have been missing. Knowing the age of the child when the picture was taken, and using data that describe the relative rates of growth of different parts of the head and face, a picture can be adjusted to compensate for these changes and show the way the child's face would be expected to appear at a different age.

The process is carried out by locating a series of reference points on the original photograph of the child's face. These points form the vertices of a series of triangles that cover the entire head with a mesh. From measurements on growing children, the relative growth and change in shape of these triangles as a function of age is known. Stretching the original image within the various triangles according to these proportions produces a warped or "morphed" image that models the age progression of an actual person. In addition, pictures of relatives may be used to capture features that are merged with the morphed image to apply shading. These pictures can be very helpful in identifying missing children, and the extent of the match to the actual child can be astonishingly close.

When working with photographs of adults, the age progression specialist mainly uses artistic skills with software to manipulate the mature features to reflect the aging process. Figure 33 shows a typical example. In this criminal case, a man had been missing for 22 years and was wanted by the FBI for "violations of Unlawful Flight to Avoid Prosecution for the crime of Murder." The photo of the suspect at age 36 (Figure 33a) was scanned into the computer, cropped, and vertically straightened. Using the stretching tool, eyebrows, eyelids, and mouth corners were pulled downward. The appearance of bagging and wrinkling under the eyes was accomplished with an airbrush technique combined with contrast adjustment. Using this same technique, cheek folds and brow furrows were deepened. Additionally, the neck was aged appropriately. Finally, an eraser tool was used to crop and smooth the hairstyle for an updated look. The result (Figure 33b) shows the projected appearance of the suspect. With the assistance of this image, the suspect was arrested in 1997 by the FBI after eluding them for almost 24 years. A photograph of the suspect at age 58 (Figure 33c) shows many points of similarity to the projection.

Similar techniques are used to alter the appearance of photographs to show changes in hair styles and other cosmetic changes that may be used by someone wishing to change their appearance from existing photographs.

Figure 33 Age progression of a suspect's face: (a) photograph at age 36; (b) age-progressed image as discussed in the text; (c) photograph of the suspect when apprehended at age 58 (courtesy Dr. Mary Manhein, Louisiana State University FACES Laboratory).

Another application for these methods, also involving physical models, is the reconstruction of the appearance of a face based on remains, by building up the contours of facial muscles. Figure 34 shows an example. Examples of the techniques used for age progression, facial reconstruction, and composite drawing, and the various kinds of results obtained, can be found in K. T. Taylor, 2000, *Forensic Art and Illustration*, CRC Press, Boca Raton, Florida, ISBN 0-8493-8118-5.

Within the general topic of creating and modifying images of faces is the generation of "police sketches" of criminals. This process is now sometimes performed with computer software that allows a small set of "standard" facial components (eyes, mouth, nose, ears, etc.) to be positioned, stretched, or compressed to form an image of a perpetrator (Figure 35). After appropriate tonal shading, the resulting composite can be printed or distributed electronically. In the examples in Figure 36, computer software has been used to generate images of several famous people. Even though actual photographs were available for reference in the process, the similarities generally are restricted to a few relatively trivial characteristics such as prominent sideburns or unruly hair.

Such software is very easy to use for generating generic face images, but is unable to reproduce such unique details as an unusual ear or nose, a twisted lip or facial asymmetry, a noteworthy skin condition or scar, or even a facial expression, which might be the most strongly remembered discriminating features seen by the witness. A skilled police sketch artist can capture these details, as shown in Figure 37, and in fact the most important skill such artists possess is the ability to elicit those details from the witness. In this case, the only role of the computer is to digitize and distribute the sketch.

Figure 34 Reconstruction of the face of a white male, approximately 50 years of age, based on remains (courtesy Dr. Mary Manhein, Louisiana State University FACES Laboratory).

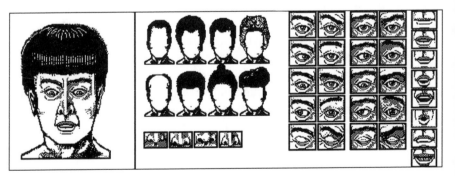

Figure 35 Assembling features to create a computer-generated sketch of a face.

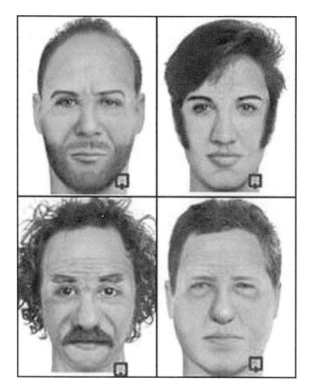

Figure 36 Images of famous faces generated using computer software: Bruce Willis, Elvis Presley, Albert Einstein, and Bill Clinton. (Courtesy Interquest FACES, Quebec, Canada.)

Of course, it is always important to remember that since these pictures are based on the recollections of witnesses, they are notoriously unreliable as accurate depictions of the actual persons. As discussed in Chapter 4, recognition and identification of faces of unfamiliar people, particularly ones from different racial or ethnic groups, is a complicated process. The sketches in Figure 37 show that while eyewitness information does report some features of each individual, there are many other details that are not reported or are reported incorrectly. The sketches share some characteristics with the actual mug shots, but would not necessarily enable someone else to recognize the person.

Figure 37 Police artist sketches showing subtlety of detail not available with software-generated faces, with comparison mug shots (beneath) of each individual taken after his subsequent apprehension. (Courtesy Gil Zamora, San Jose, California Police Department.)

Enhancement

3

All of the techniques for noise reduction, adjustment of tonal range (grey scale brightness and contrast) or color balance, etc., discussed in Chapter 2 can be considered as enhancement of the original image in that they improve the ability of the human vision system to view the image and extract important information from it. But, because these are, for the most part, operations that were routinely accomplished by accepted darkroom techniques before digital imaging and computers became available, they are not usually considered controversial. The more aggressive methods of image enhancement surveyed in this chapter may raise some issues of acceptability because they are not so easily accomplished without the computer (although in principle many of them can be performed using darkroom equipment and film).

The general topic of enhancement is one of the most controversial areas of processing of digital images. Many people feel that these operations "create" information, and that if some detail isn't readily visible in the original photograph it is improper to use the computer to make it evident to the viewer. Certainly there are some methods that can result in alteration of the actual information present, and these are explained carefully below. But first it may be helpful to discuss some techniques that are definitely appropriate, and explain why this is so.

Enhancement of Detail

Human vision is a very interesting subject. One of the things that researchers have learned is that very little of the content of a typical scene is actually passed up through the retina, optic nerve, and optical cortex to the areas of the brain that deal with the information. Most of the static parts of the scene, and most of the fine detail, is ignored. In routine living, it is only when some event causes this information to become important that we go back and examine it. So the idea that details will automatically be noticed is simply not true.

Also, the retina of the eye works to extract edges and corners of features and creates what amounts to a sketch of the scene, and it is these outlines that are transmitted to the brain. This is why cartoons, which are similar outline sketches, are so effective at communication. So anything we can do to help the eye locate and extract these edge details will be helpful. That is why changing the illumination of a scene, for example by using low angle incident lighting to highlight surface scratches, is useful. We have all had the experience that details that do not show up under one lighting condition become evident in another. A trivial example is dust on a table that becomes very obvious when strong sunlight strikes it at a low angle.

Altering a light source to create edge lighting or backlighting is not "creating" any new information, it simply helps the rather limited functioning of the human visual system, which ignores uniform areas and notices sudden variations, to extract information from a scene that might otherwise be overlooked or ignored. Likewise, there are image processing steps that can assist the human viewer in noticing details that have been captured in the recorded image (on either traditional film or a digital camera), but would be overlooked without enhancement.

By no means is this restricted to digital image processing in a computer. Many of these methods have deep roots in traditional photography and darkroom processing. One of the most useful methods, which dates back at least 100 years (I have a 1920 Kodak Darkroom Manual that treats it as a standard method) is unsharp masking. This was used particularly by astronomers, who have images containing very bright objects (stars, galaxies, dust clouds, and other structures) as well as very dark regions (empty space) in their photographs. The dynamic range of film is barely great enough to capture both the bright and dark information, but the human viewer cannot see the details in both the bright and dark areas at the same time. Prints can be made with adjusted contrast and gamma to show one or the other, but not both.

Unsharp masking solves this problem. As illustrated in Figure 1, a print is made from the original negative onto another piece of film, the same size as the original, but out of focus. This film is developed and is a positive transparency. Superimposing the positive and negative transparencies and printing through both of them to make a print produces the final result. In areas where the original negative is very thin (little exposure), the positive transparency is dense and the result lets little light through. Similarly, where the original negative is very dense, the positive transparency is thin and the overall amount of light that penetrates to the final print is also low. But where the original negative has fine detail, the positive transparency is not in sharp

Figure 1 The process of unsharp masking: (a) original image with high contrast and fine detail; (b) blurred print with inverted contrast; (c) combining (a) and (b), the composite print has overall contrast suppressed and fine detail enhanced; (d) adding (c) back to (a) to superimpose the enhanced fine detail onto the original image contrast.

focus and cannot match the variations in the original. The result is to let the fine detail through, and produce an image in which the detail is visible. Side-by-side comparison of the final result with the original reveals that all of the fine details that have been enhanced are actually present in the original, although they might not have been visually noticed there.

Sharpening

This same process, often called by the same name (unsharp masking), is carried out straightforwardly on the computer. The process of Gaussian blurring discussed in Chapter 2 produces the same effect as an out of focus print, and the amount of defocus is adjustable. Subtracting the blurred image from the original can also be carried out with an adjustment for the amount of subtraction. The result is an ability to enhance fine detail that is present in the original image, while suppressing the overall contrast present.

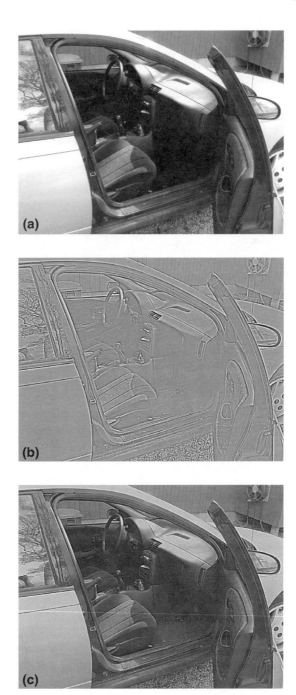

Figure 2 Unsharp masking on the computer: (a) original high contrast image (identical to Figure 14 in Chapter 2); (b) unsharp mask applied; (c) sum of (a) and (b) to superimpose fine detail onto original image contrast.

In the example shown in Figure 2, a copy of the original image was subjected to a Gaussian blur with a standard deviation of 10 pixels and then subtracted from the original. The principal objection to this result, although it shows fine detail present in the original image that is not readily visible without the processing, is that the suppression of the overall brightness variation makes the image more difficult to interpret. The usual solution to this difficulty is to add the detail from the unsharp mask back to the original image, producing a result in which edge detail is enhanced by having a greater local contrast while the overall variation of brightness is still present.

This process is generally called a sharpening operation. It is one example of a range of related operations that can be applied. The simplest of these is to convolve the image with a neighborhood of kernel values, such as:

$$\begin{array}{ccc} -1 & -1 & -1 \\ -1 & +8 & -1 \\ -1 & -1 & -1 \end{array}$$

This instructs the computer to multiply the numeric value of the stored brightness in each pixel by 8 and to subtract the 8 neighboring pixel values from it. In a uniform area of the image this would produce a zero result, so a medium grey value (128 for a typical 8 bit image) is added to the result. If any of the neighboring pixels is brighter or darker than the central one, the result will be darker or brighter than medium grey, respectively. This process, called a Laplacian, acts like an unsharp mask but with a fixed and very small amount of blurring. The subtraction of the average of the immediate neighboring pixel values is the blurred copy of the image. Because the Laplacian also eliminates the overall brightness variation in the image, it is again desirable to add the result back to the original image. This can be done simply by changing the central pixel weight from 8 to 9 (in which case there is no need to add 128 to the result). Figure 3 shows an example.

This is a classic sharpening filter that enhances edge detail in an image. It is much less flexible than the unsharp mask, because of the fixed small size of the blur. A more general process using the unsharp mask approach is to make a duplicate of the original image, blur it enough to eliminate the fine detail of interest, and then subtract that from the original. The difference between the two is just the detail, which is then visible. Notice that all of these operations alter a step in brightness that marks the edge of some feature or detail in the image, as shown in Figure 4. One consequence of this type of processing is that it alters the relative brightness of different regions in the image, so that they cannot properly be compared. This illustrates the need to perform some types of comparison and analysis on the original images, and to use processing to extract other types of information from the images.

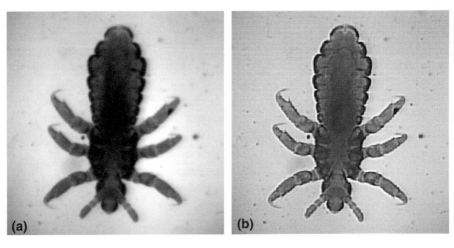

Figure 3 A one-step sharpening operation using a 3 × 3 kernel of weights: (a) original; (b) result.

Both the Laplacian and the more general unsharp mask have a problem with images that contain random noise. The noise is, as discussed in Chapter 2, a variation in brightness between one pixel and its neighbors in regions that should be uniform. The sharpening process increases that difference, making the noise more visible. In a noisy image, it is therefore essential that the noise be removed before the sharpening process can be used, or the increased visibility of the noise will eliminate any advantage which sharpening might bring to the visibility of the important detail.

Figure 5 shows the effect of noise present in the original image. The noise is enhanced more than the detail, and it is actually more difficult to read the license plate number after enhancement (Figure 5b) than in the original because of the visual distraction.

The most general form of this type of sharpening enhancement, the difference of Gaussians operation, is able to deal with random image noise at the same time that sharpening is performed. The procedure is to make two copies of the original image. Both are smoothed using a Gaussian smoothing operation. The first copy is smoothed using a small standard deviation that removes the random noise but not the details of interest. The second copy is smoothed using a larger standard deviation, which removes both the details and the noise. The difference between these two images is then able to show just the details. In the example of Figure 5c, the noise was removed with a Gaussian smoothing using a standard deviation of 0.33 pixels, and the detail was removed with a Gaussian smoothing using a standard deviation of 3 pixels. Typically the ratio of the two standard deviations is between 4 and 12, but as this is a visual enhancement function the optimum

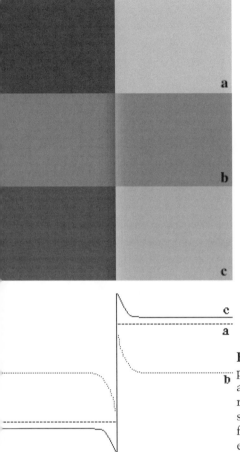

Figure 4 Diagram of the sharpening process with brightness profiles across a step: (a) original step in brightness; (b) result of Laplacian or unsharp mask, suppressing the average brightness difference but leaving bright and dark edges adjacent to the step; (c) sharpening result by adding (b) to (a).

setting for any particular image is selected interactively in order to produce the best result.

All of these sharpening methods are called "high pass" filters because they remove gradually varying (or in electrical engineering terms "low frequency") changes in brightness values while preserving the rapidly varying ("high frequency") changes. The methods have been described here in terms of the actual image pixels, because that is familiar to most people dealing with digital images. In a later section in this chapter, the application of frequency space (Fourier space) methods is introduced. The same low pass (Gaussian smoothing) and high pass (sharpening) operations can be performed using those methods, which are sometimes faster in implementation. Mathematically, convolution operations in the spatial domain (pixels) and the frequency domain can be shown to be identical. This is not true for

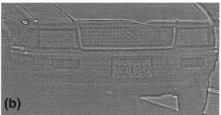

Figure 5 Effect of noise on sharpening: (a) original image, with just-visible random noise speckle; (b) application of a sharpening operation increases the visibility of the noise and decreases the visibility of the information; (c) application of a difference of Gaussians operation, reducing the effect of the noise and sharpening the detail.

ranking operations on pixels (such as the median filter discussed in Chapter 2), which have no frequency domain equivalent.

Sharpening Color Images

Unlike the example shown in Chapter 2 of smoothing a color image by processing the red, green, and blue channels, sharpening operations generally require converting the image to a different color space. If a sharpening operation is applied to the red, green, and blue channels of a color image, the results are generally very poor. The numerical value of each color is changed according to the difference between it and the neighbor values, and because of the noise this causes different value changes in each color channel. When these are displayed, the result is to shift the hue (color) of the pixel. Sharpening in this way produces a speckled appearance with pixels having colors that were not present in the original scene and which are very visually distracting. Figure 6 shows an example (the same noisy color image used in Chapter 2) showing the results of applying a sharpening filter to the RGB planes, compared to processing the intensity only (with an unsharp mask) while leaving the color information unchanged.

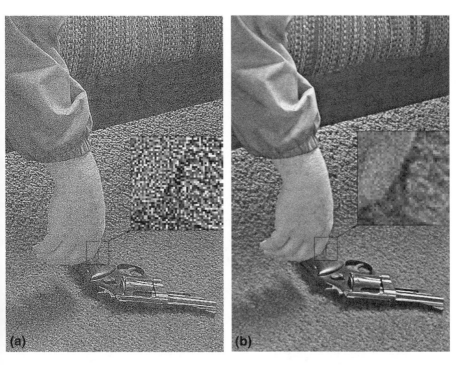

Figure 6 Sharpening a noisy color image (from Chapter 2, Figure 8): (a) sharpening the individual red, green, and blue channels (shown as Plate 15); (b) unsharp mask (standard deviation 5 pixels) applied to the intensity only (shown as Plate 16).

This requires a brief digression into color space terminology. As shown in Figure 7, the RGB color space in which the cameras collect information and computers store it, has a simple orthogonal or Cartesian coordinate system. Any color that the system can represent is present as a mixture of red, green, and blue. This is convenient for the hardware, but is not a good model for how humans perceive color. Few people, even those with experience in color science, look at an orange (for example) and can determine the amount of red, green, and blue present.

The color space used by most artists and those concerned with color presentation and matching is based on a coordinate system using hue, saturation, and intensity (HSI). There are several variants of this space, whose differences are not important for our purposes here. Generally, the vertical axis of this space is the intensity (sometimes called brightness or value), which is the intensity that would be recorded by a grey scale camera or "black and white" film that was not sensitive to color. The introduction of color moves points away from this axis. The direction of the shift is the hue, which corresponds to what most people mean by color. The color wheel that

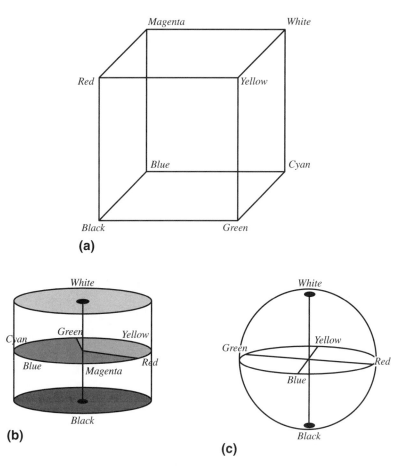

Figure 7 Several color space coordinate systems: (a) RGB coordinates; (b) one form of HSI coordinates; (c) L*a*b coordinates.

children learn in kindergarten, progressing as shown in Figure 7 from red to orange, yellow, green, cyan, blue, magenta and back to red, represents a variation of hue. The amount of color present is the saturation, represented as the distance from the central grey scale axis. The difference between pink and red, which have the same hue, is one of saturation. Likewise the difference between sky blue (unsaturated) and deep ocean blue (more saturated).

These three values can be combined to represent any color. The mathematical details of the color space, which may be represented as a cylinder, a cone, a hex-cone, or a double cone, are not particularly important for the purposes of image processing. These various versions of color space are called HSV (V for value), HSB (B for brightness), or IHS. There is even a spherical color space (L*a*b) in which the north–south axis is the brightness (**L** for Luminance), and instead of the radial coordinate system, two orthogonal axes (**a** and **b**) are used to represent the red-green and yellow-blue variations.

This coordinate system offers some arithmetic simplifications for efficient computation. What matters is that the RGB components can be converted to HSI ones using a simple set of equations or a few lines of computer code.

Once converted to an HSI space, the processing is performed on the grey scale intensity values and the color information is not changed. Then when the image is converted back to RGB space so that it can be displayed, there are no shifts in color or new colors added to the image, as shown in Figure 6b. In many systems, this conversion is performed automatically when processing operations are performed on color images. After processing, there may be still be more noise present in the color information than in the brightness signal, but the human visual system is very tolerant of this and depends far more on the brightness information to delineate the structure of scenes. This is why broadcast television, for example, allocates less bandwidth for the transmission of color data than for brightness information.

It is also possible to perform a median filter operation directly in color space. As mentioned in Chapter 2, this requires a new definition of the concept of a median. For brightness values, it is straightforward to rank the pixels into order from brightest to darkest and select the central value in the list. This is the usual statistical meaning of "median value." For pixels with color information, whether RGB or HSI, each color pixel requires three numerical values and is plotted as a point in three-dimensional space. In this case, the median value is defined as that point which lies closest to all of its neighbors. The distances from each pixel's coordinates to the coordinates of all of the other pixels in the neighborhood are summed, and the pixel with the lowest total is taken as the median. In the Cartesian RGB space, the distances are calculated using the Pythagorean relationship; in cylindrical or conical HSI spaces the calculation is more complicated, and in particular the hue values must be taken in the shortest direction (modulo arithmetic). The color values for the coordinate that meets this definition of median are then assigned to the central pixel in the neighborhood in the derived image. Figure 8 in Chapter 2 shows an example. Obviously, this requires more computation than the methods described above, but still takes only a few seconds for typical images.

Other Enhancement Techniques

The high pass filters discussed above increase the visibility of detail by magnifying local changes in brightness while suppressing the overall range of contrast. Another approach to accomplishing a similar result is local equalization. To explain this, it is necessary to first go back to the topic of manipulating the contrast of the image discussed in Chapter 2. In the examples shown there, the image brightness, contrast, and gamma were adjusted to improve visibility of details in bright or dark areas of the image. These types of operations are global in nature: any two (or more) pixels that start out with the same brightness will be altered in the same way, to some new brightness value, without regard to the neighborhood in which they reside. Furthermore, the relationship between the original brightness value and the final adjusted value follows a smooth function, either a straight line or a logarithmic curve.

There is another type of relationship that can be created based on the actual contents of the image. Called equalization, this attempts to distribute the intensity values in the final image as uniformly as possible across the range of possible display brightness values. It is accomplished by using the image histogram. The usual histogram is shown as a plot of the number of pixels as a function of intensity. The integral of this curve is the plot of the number of pixels that have all brightness values equal to or less than each value of intensity. Figure 8 shows an image with its conventional histogram and the cumulative histogram. If the cumulative histogram is used as the function that relates the original to the adjusted brightness, the result is an image in which the cumulative histogram becomes a straight line, as shown. The conventional histogram then shows gaps, which result from spreading out the grey values in the peaks of the original histogram. The result in terms of the image itself is to make slight variations of brightness more evident in both predominant light and dark areas.

This method for modifying image brightness is often used in laboratory situations in order to make it easier to compare images taken from different specimens or under different conditions, or to reveal gradual changes in brightness that are otherwise difficult to see (human vision is relatively insensitive to gradual changes in brightness). However, histogram equalization is rarely useful for real-world scenes because it alters the contrast in ways that are unfamiliar to the viewer and make it more difficult to recognize familiar objects.

Local equalization uses the principal of histogram equalization, but applies it to a neighborhood. If a small region (typically 3 to 15 pixels in diameter) is equalized, the result will be new intensity values for all pixels. But only the new value for the central pixel is saved and used to build up a new image, pixel by pixel, as the neighborhood is shifted to all possible positions in the original image. The consequence of local equalization is to make pixels that are slightly darker than their surroundings darker still, and conversely to make pixels that are slightly brighter than their surroundings brighter still. As shown in Figure 9, this has the effect of making patterns such as fingerprints more visible, particularly when they are superimposed on a background that varies in brightness. Of course, the first steps that must be taken in recording fingerprints are

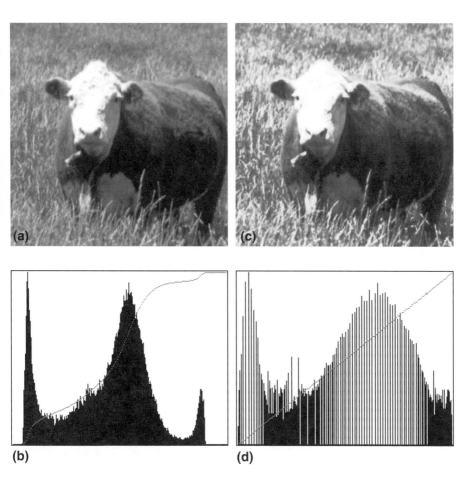

Figure 8 Equalization: (a) original grey scale image; (b) histogram of (a), with cumulative histogram superimposed as a grey line; (c) image after equalization; (d) histogram of (c), with cumulative histogram superimposed as a grey line.

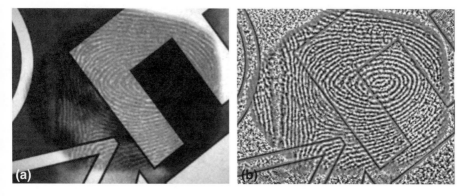

Figure 9 Fingerprint on a magazine cover: (a) original image; (b) after local histogram equalization with a 9 pixel wide neighborhood.

to acquire the images under the best possible conditions, which typically means low angle lighting (or in some cases backlighting if the fingerprint is on a glass or other transparent material).

This type of local equalization, like all of the other techniques discussed so far in Chapter 2 and this chapter, has been used for many years, and is considered a standard technique that is covered in most introductory books on image processing and implemented in most computer programs. It has the same sensitivity to noise that the unsharp mask and Laplacian do, because it amplifies variations between the central pixel and its surroundings. Comparatively recent developments have broadened the applicability of the local equalization method (e.g., see T. Gillespy, 1998, "Algorithm for Adaptive Histogram Equalization," *Proc. SPIE*, Vol. 3338, p. 1052–1055). Termed "adaptive local equalization," the method allows for noise reduction by using the median intensity value in the neighborhood, and allows weighting the pixels in the neighborhood according to their distance from the central pixel. Also, like the sharpening filters discussed above, it allows adding back the original image to the processed result.

Adaptive equalization is less noise sensitive and reveals local detail and variations in both the dark and light regions of the original image; Figures 10 and 11 show examples. When applied to color images, the RGB data are converted to an HSI space, the intensity values are modified, and then the result is converted back to RGB for display; Figure 12 shows an example.

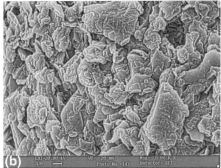

Figure 10 Scanning electron microscope (SEM) image of dried paint: (a) original, with a high contrast due to bright regions on sloping surfaces and a dark hole; (b) after adaptive equalization, showing fine detail in both the bright and dark areas. Note that all of the detail thus revealed can actually be found on the original, once it is made evident by the processing, but is likely to be overlooked in the original.

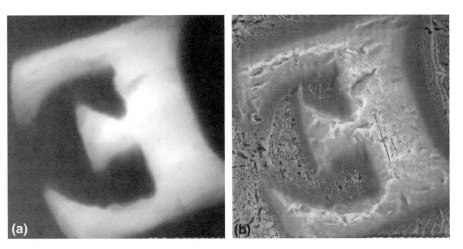

Figure 11 Surface markings on a coin, imaged by scanned probe microscopy: (a) original, with high contrast due to the relief of the lettering on the coin; (b) after adaptive equalization, showing fine detail and scratches in both the bright and dark areas.

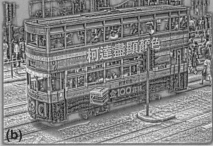

Figure 12 Color image: (a) original, with high contrast and little visible detail in shadow areas; (b) after adaptive equalization, showing enhanced detail and improved visibility in shadow areas (see Plates 17 and 18).

In many cases, the use of adaptive equalization is superior to a traditional high pass filter for enhancing the visibility of fine detail, such as the firing pin and extractor marks on the shell casing in Figure 13.

A word of caution about any processing algorithms: their application should always be to the entire scene in the image and not to a region or object within it. Human vision is comparative by nature, and the viewer of an image expects to be able to compare objects or regions within an image. If one area has been altered by processing, for instance to sharpen edges or alter contrast, subsequent interpretation of the image can be affected. Such alterations give rise to many of the concerns about the effects of "tampering"

Figure 13 Closeup view of the back of a 32 caliber shell casing: (a) original image; (b) unsharp mask; (c) difference of Gaussians; (d) adaptive equalization.

with digitized images. The same restrictions should apply to the film photographer, who would not be justified in using comparable techniques in the darkroom to enhance (or reduce) the visibility of detail in only one part of an image. Making adjustments to an entire image to enhance the visibility of detail still allows the viewer to interpret the image correctly by providing correct comparative references to other objects within the scene.

While extremely useful for revealing detail such as the examples shown here, local equalization and high pass filtering are not generally helpful in processing images of faces, because they alter the contrast in ways that conflict with the normal human ability to recognize facial features. These methods may be useful, however, for revealing scars or tattoos, especially on dark skin.

Less Common Processing Methods

There are other techniques for image processing that are often applied in specific situations such as machine vision (recognizing and/or measuring objects in industrial situations) and other areas of image processing, which are generally not useful for forensic imaging applications that use image processing as an aid to visual interpretation because they either alter the image contrast in ways that confuse normal visual recognition, or because they can remove details from images. They may be useful and appropriate in specific situations, but their use will always need to survive *Frye* or *Daubert* challenges (as discussed in Chapter 5) and the expert using them will need to be careful to show how and why the methods were used in each circumstance.

One technique, inherited from darkroom procedures on film, is called solarization. It reverses the normal contrast in either the bright or dark regions of the image so that increased contrast can be used there to reveal fine detail. In the darkroom, this was accomplished by a series of chemical treatments and exposure to light. In the computer, it is easily performed (perhaps too easily, inviting misuse) by constructing a curve of displayed brightness vs. stored brightness that is multiple-valued. In other words, in the final image there are features shown with the same brightness that were originally very different, which conflicts with our normal experience. However, as shown in Figure 14b, this can reveal structures that would otherwise be difficult to discern.

In classical image processing texts, there is usually considerable attention given to edge-finding methods. These highlight boundaries of objects and features, and are used to delineate them for measurement or to construct logical representations of objects for machine vision computation. There are a wide variety of methods used to compute these, but the results are generally rather similar. The example shown in Figure 14c uses one of the most common implementations of this method, the Sobel operator. This calculates the magnitude of the maximum derivative of brightness in any direction at each point in the image and creates a new image in which the pixel values represent this magnitude. The calculation is performed by calculating the derivatives in the horizontal and vertical directions, and combining them as vectors (square root of the sum of squares of the two magnitudes).

Another operation that is used to remove features and detail from images is the top hat operator. This is designed to find structures of a particular size and shape and to ignore anything that is larger or smaller. In Figure 14d, the top hat was set to find dirt particles with diameters up to 7 pixels. These are shown in the final result, but the principal structure present (the insect) has been eliminated completely. The top hat operator is a specific example of erosion-dilation operations that remove or add pixels to features. Generally,

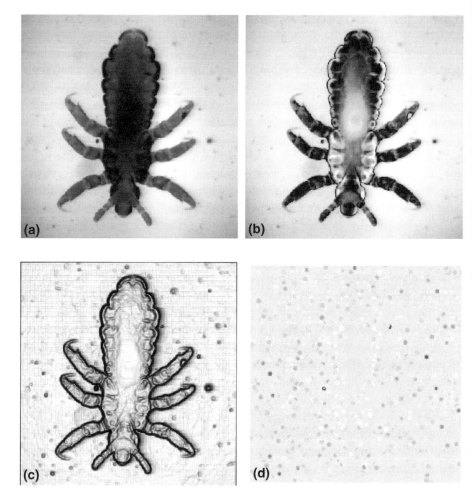

Figure 14 Microscope image of a head louse: (a) original image; (b) solarization, expanding the contrast in the legs and reversing the contrast in the darker body so that it is shown as a photographic negative; (c) Sobel edge operator, which delineates all of the internal and external boundaries in the image; (d) top-hat filter, which eliminates everything except features (in this example, dirt particles) with a specified size and shape.

these are inappropriate for forensic work because they alter the size and shape of structures, including removing them entirely from the image.

Color Separation and Filtering

With conventional film-based photography, it is common to use colored filters to selectively increase image contrast by absorbing light whose color

Figure 15 Color filtering: (a) original color image (see Plate 25); (b) the red channel only; (c) applying a green-cyan filter.

is complementary to the filter color (complementary on the familiar color wheel: red is complementary to cyan, yellow to blue and magenta to green). Thus a yellow filter will darken blue sky in a photograph, increasing the contrast of clouds. This is most often done when the photograph is taken, and of course the same technique can be applied to digital cameras, with the same results. It is also possible to use colored filters in the darkroom, when making prints from colored negatives.

The analogous computational technique with digital cameras applies the equivalent color filter to the stored image. A vector representing the filter color (as a combination of red, green, and blue) is combined with the color vector for each pixel to form the dot product that represents the amount of light that would pass through the filter. This becomes the intensity value for the pixel. For the particular case of red, green, or blue filters it is possible to simply use the corresponding stored R, G, or B intensity.

In the example shown in Figure 15, the red channel shows the lily without any spots, because the red spots have the same amount of red as the white petals. But applying a green-cyan filter (selected in this instance with a wavelength of 520 nm) absorbs red and produces strong contrast for the spots. This type of complementary color filtering is easily accomplished in the computer, because a filter of any color can be created as needed, to produce results similar to traditional film photography.

Photography is sometimes used in conjunction with the examination of documents, for example to show magnified portions of the document to

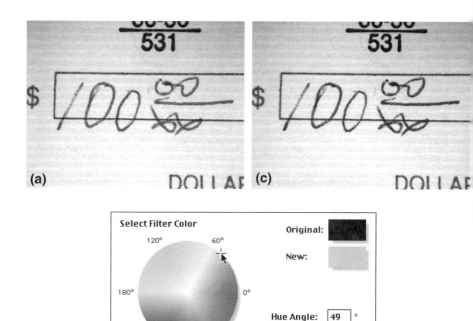

Figure 16 Color filtering: (a) original image; (b) selection of a yellow filter complementary to the blue lines; (c) filtered image (Plates 19, 20, and 21).

compare lines, handwriting or printing styles, paper watermarks, etc. The use of digital imaging does not alter these applications at all, and because of the flexibility of manipulating the color image, may offer some efficiency or flexibility. For example, Figure 16 shows a macro image of handwriting on a check. The zero that converts the amount from 10 to 100 dollars has been written with a different pen. In the original color image, the difference between the two blues is not visually evident. Applying a yellow filter (complementary to the blue) leaves the original pen lines perfectly grey, but shows clearly the greenish cast of the imperfectly matched second pen. This can be done, of course, with a physical filter, but the ability to select any color filter with the computer simplifies the process.

Color separations in the computer are also possible in other spaces. These are often more powerful than those performed in RGB space and not possible with physical filters and photographic film. It was pointed out before that human vision distinguishes colors based on hue (color), saturation (amount of color), and intensity (brightness). Separations performed in this space generally correspond better to the color distinctions that people visually recognize in images. For example, in Figure 17, the difference between the brown

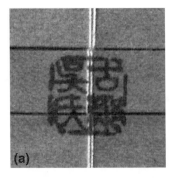

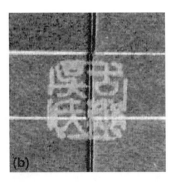

(c)

Figure 17 Separation in HSI space: (a) original color image (shown as Plate 22); (b) product of saturation and intensity values for each pixel; (c) selected pixels with characteristic hue, saturation, and intensity.

background and the red chop mark (printed signature) is subtle. Brown is essentially a darker and less saturated red-orange, not easily distinguished from red. Converting the image into an HSI space and selecting pixels based on these characteristics permits isolating the signature as shown.

This approach can be extended to even more difficult situations. In the example shown in Figure 18, the original image has a dye-enhanced (magenta) fingerprint superimposed on a bluish background with dark blue and black lettering. Converting the image to HSI space enables a rather unusual use of the contrast expansion technique shown in Chapter 2. Expanding the range of the hue, saturation, and intensity values (as shown in Figure 18b) and then transforming the image back into RGB color for display produces a bizarrely colored image (Figure 18c) in which the HSI color differences between the various features in the image have been maximized.

This image was then converted to a grey scale image (Figure 18d) using a regression technique (shown schematically in Figure 19). Each pixel's color coordinates correspond to one point in color space. For all of the points representing the entire image, there is a straight line that gives the best fit (the points have the least scatter about that line). Fitting this line uses standard mathematical regression techniques. This line is then used to construct a new grey scale image in which the position of each pixel's coordinates along that line becomes the pixel intensity. This "optimal grey" transformation produces a grey scale image in which the fingerprint is clearly seen without the written characters evident in the original picture.

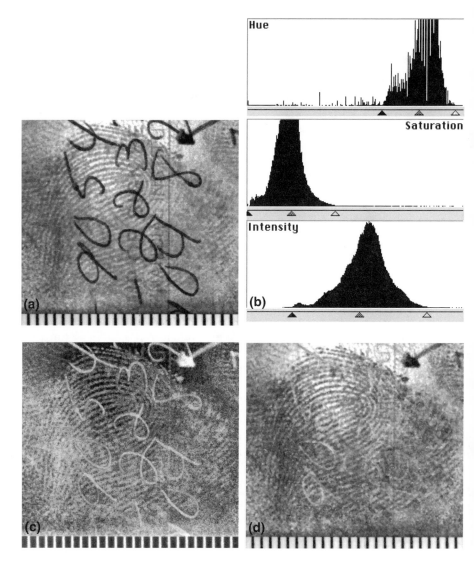

Figure 18 Expansion of the HSI range: (a) original image (shown as Plate 23; courtesy of George Reis, Imaging Forensics, Fountain Valley, California); (b) histograms for the H, S, and I channels with limits set to expand their range; (c) resulting image with maximum color contrast (see Plate 24); (d) optimal grey image derived from (c).

Frequency Space (Fourier Processing)

Most classic image analysis texts devote considerable coverage to the processing of images in frequency space. This is partially due to the fact that there are some types of processing, such as the removal of periodic noise and blur,

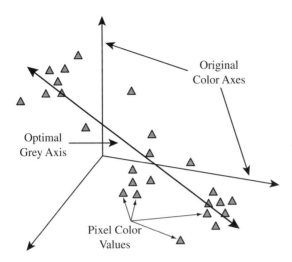

Figure 19 Principle of using regression to find an optimal grey scale axis. Each point represents the color coordinates of one pixel in the image. The best fit line through the points in color space defines the new grey scale axis, along which the projected position of each pixel's coordinates defines a grey scale brightness.

that can only be understood in those terms and performed in that way. It is also due in no small part to the fact that the mathematics is familiar to electronics engineers and physicists, who deal with processing images in much the same way that they process other signals. Rather than introduce any of the mathematics here, the emphasis will be to show by example the relationship between the spatial or pixel domain (the original and familiar image) and the frequency space representation.

Joseph Fourier showed that it was possible to form any signal waveform (in our case, we are interested in the function of image brightness as a function of position) as a summation of sinusoids of increasing frequency. The practical use of this method for decomposing the signal into the sinusoidal components, analyzing or altering the amplitudes of the terms, and recombining them to make a processed signal (in our case, a processed image) has been one of the mainstays of electronic engineering and (more recently) computer-based signal processing for most of the past century. As shown in Figure 20 for a simple case, any image can be converted, using well understood and widely implemented algorithms, from the spatial domain to the frequency domain, and back again with no loss of information. In the frequency domain image, the amplitude of each sinusoidal component is represented by the intensity of a point whose distance from the center represents the frequency of the sinusoid (the inverse of its wavelength, or the spacing of the wave crests) and whose direction from the center represents the orientation of the sinusoids. This display is called the "power spectrum" because

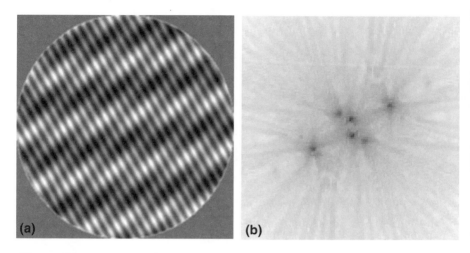

Figure 20 An image and its Fourier transform power spectrum: (a) a set of three overlapping sinusoids with different orientations and frequencies; (b) the power spectrum (enlarged), showing three points corresponding to the individual sinusoids. The darkness of each point represents the amplitude of that term. Note that the conventional way of displaying the power spectrum repeats the same information in the bottom half as in the top half, so the three points are each plotted twice, symmetrically with respect to the center or origin.

it represents the amount of energy associated with each component sinusoid. There is additional information, the phase or offset of each sinusoid, that is usually not displayed but is also needed to reconstruct the original image.

The actual computation of the frequency space representation of the image is usually performed with a "Fast Fourier Transform" (FFT), a well-known algorithm that has made it practical to implement this technique with modern computers. In the most common implementation of this algorithm the image size must be exactly a power of 2 (128, 256, 512, 1024, etc.) pixels in dimension; for irregularly sized images the usual solution is to pad them out to the next larger exact size. There are other transforms that can also be used, based on other sets of functions than sinusoids (called the "basis functions" of the transform). The JPEG algorithm widely used for image compression is based on a cosine transform, closely related to the Fourier algorithm; it is applied to the hue, saturation, and intensity information taken from the original image. Wavelet transforms using a variety of basis functions are also used for compression; they have an advantage over the Fourier transform in dealing with the edges of images, but are generally not so easily interpreted nor used for filtering out periodic noise or enhancing detail.

High and Low Pass Filtering, Bandpass Filtering, and Periodic Noise

For a typical complex image, the Fourier transform representation does not directly reveal much information. Typically the amplitude of the power spectrum drops off with distance away from the center (higher frequencies have less amplitude). As shown in Figure 21, the presence of a vertical and/or horizontal line on the power spectrum results when the left and right edges, or the top and bottom edges, of the picture do not match.

Altering the amplitude of the terms in the Fourier representation of the image can be done simply by multiplying the values by a constant. A filter like the one shown in Figure 21c attenuates the low frequencies while passing the high ones (it is generated as a simple linear ramp, uniform in all directions). If this filter is multiplied by the amplitude of the power spectrum and the image is converted back to the spatial domain (Figure 21d), the result is to suppress low frequencies (gradual variations of brightness), so this is called a high pass filter. The result looks like the results from high pass filters such as the Laplacian or unsharp mask discussed in terms of the pixel array, and indeed it can be shown mathematically that the two operations are identical. This is, in fact, the reason that this class of operations is called by the name "high pass filters" when they are performed in the spatial domain (the array of image pixels).

Similarly, a filter that suppresses the high frequencies and passes the low frequencies (Figures 21e, f) produces smoothing and reduction of random noise. A filter that suppresses both low frequencies (to reduce overall variations in brightness) and high ones (to suppress noise) is called a bandpass filter, and corresponds approximately to the difference of Gaussians discussed in Chapter 2 (the Fourier transform-based filter can have sharper frequency cutoffs than the spatial filtering technique). A bandpass filter corresponding to the spacing of friction ridges in human fingerprints can be used to sharpen the appearance of fingerprint images, by eliminating low and high frequencies and keeping just those of interest. Figure 22 shows an example.

There is no equivalent in frequency space processing to operations like the median filter, which is based on ranking of the pixel values. But any convolution that can be accomplished by multiplying the pixel values times a set of weights in the spatial domain can also be performed in the frequency domain, and is often faster to carry out there because only one multiplication per pixel is required, whereas with a large filter, many multiplications and additions are needed for each pixel in the image.

If the only use of frequency space processing was to provide faster implementation of these operations, it would probably not be important to know about it. The explanations based on the spatial domain pixels are usually

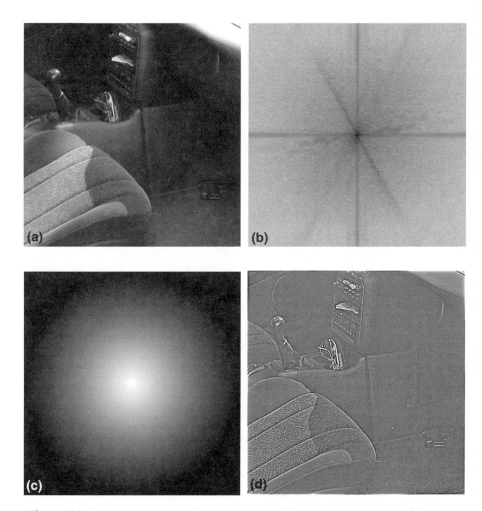

Figure 21 Filtering the power spectrum: (a) original image; (b) the Fourier power spectrum; (c) a high pass filter; (d) result of applying the high pass filter and retransforming the data to the pixel or spatial domain; (e) a low pass filter; (f) result of applying the low pass filter and retransforming.

easier to understand for those not comfortable with the mathematics of the Fourier transform, and the details of how the computer software implemented the convolution would be unimportant. However, there are several other uses of the frequency space data that are not so easily understood or implemented in terms of the pixel array.

The most important of these is the removal of periodic noise. This can result from electronic interference, moiré patterns due to printing, vibration, or other causes. It is marked in the spatial domain image by the superimposition of a regular pattern upon the image, as shown in Figure 23. In the frequency domain the one or few specific interference frequencies are

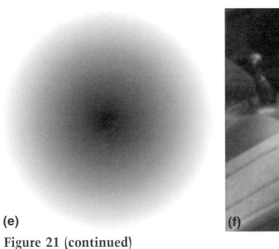

(e)

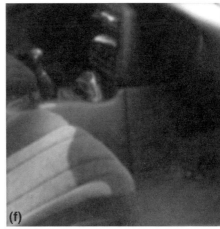

(f)

Figure 21 (continued)

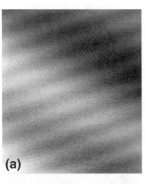

(a)

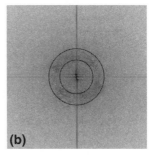

(b)

(c)

Figure 22 Isolation of a fingerprint based on frequencies: (a) original image, a backlit fingerprint on a drinking glass, with video noise; (b) the Fourier power spectrum with concentric circles marking high and low frequency limits for the spacing of ridges in the fingerprint; (c) result of selecting only information between the circles for retransformation.

revealed as "spikes," or large amplitudes that occur at a few points in the power spectrum corresponding to the frequency and amplitude of the noise. These spikes can be removed by setting the amplitude at those points to zero. This can be done manually, or by using image processing on the power spectrum to locate the points and eliminate them automatically (this latter technique uses the same top hat filter that was introduced earlier). In either

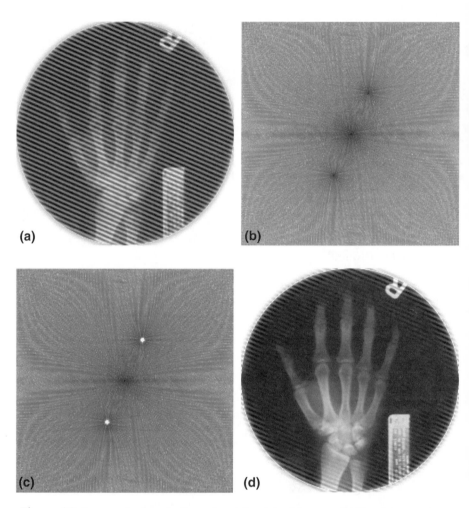

Figure 23 Removal of periodic noise: (a) original image; (b) Fourier transform power spectrum; (c) power spectrum with spike removed; (d) retransformed result.

case, the reconstruction of the spatial domain image after removal of the periodic noise spikes shows the image without the interference, and greatly improves the visibility of the image details.

This method is also very powerful as a tool for removing the moiré patterns produced by printing images using a halftone screen. If images, such as newspaper photographs, laser printer output from computers, or printed mug books, are viewed through a magnifier, the array of dots can be clearly seen. When scanned into the computer, these dots represent a major barrier to further processing. Blurring the image with a Gaussian filter can eliminate the pattern, but only at the cost of blurring important image detail as well. The fact that the dot pattern is regular allows it to be removed very effectively and efficiently in the frequency domain. Figure 24 shows an example, using

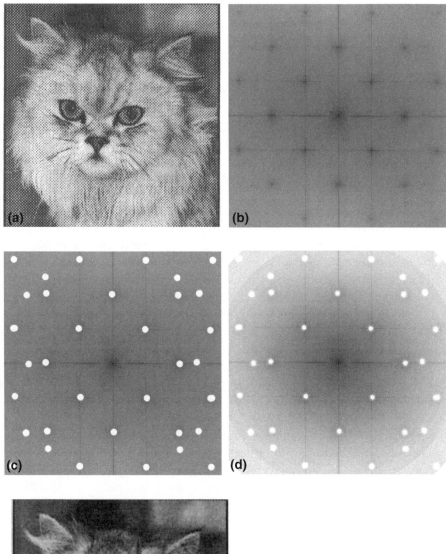

Figure 24 Removal of halftone printing pattern: (a) original newsprint image; (b) power spectrum; (c) removal of spikes; (d) low pass filter attenuates high frequencies; (e) final result.

an image scanned from a newspaper photograph. Two steps are used to produce the final image: first the regular pattern of spikes from the dot pattern is removed and then a low pass filter is employed to fill in the space between the dots. In the result, fine hairs and other detail can be seen clearly.

For color images, it is necessary to first separate them into channels for each of the inks used (typically cyan, magenta, and yellow) and then remove the patterns present for each color. These are usually in different orientations so that the dots can fit together on the page without overlapping.

Deconvolution, or Removal of System Defects

The frequency domain representation of an image is also used to remove blur due to motion or out-of-focus optics. This is also discussed in the next chapter, as a means to improve video images. But it is equally applicable to digital still images, and has been used with great success for purposes such as sharpening the images from the Hubble space telescope before the incorrectly curved optics were corrected by the installation of a new secondary mirror.

The principle of this deblurring is easy enough to understand, but sometimes very difficult to accomplish in practice. It was shown above that blurring an image by smoothing with a convolution in the spatial domain was equivalent to multiplying the power spectrum times a weighting function, which is in fact the Fourier transform of the kernel used for the spatial domain convolution. This means that if the camera system used to record the image does not capture a perfect image, due to any combination of causes, such as motion, optics, camera limitations, etc., the blurring of the image must also correspond to multiplication of the ideal power spectrum that would represent the ideal image times a filter that represents the system defects. Sometimes, as in the case of the space telescope, it is possible to calculate this filter from first principles. In almost all practical cases, it is not. But it may still be possible to measure it directly.

For astronomical pictures, this is quite simple. An isolated star should appear in the image as an ideal point. Any optical or other imaging defects will cause the star to be blurred. If the isolated image of a blurred star is captured (often called the system "point spread function") and subjected to a Fourier transform, the result is the filter that was applied to the ideal power spectrum from the theoretically perfect image. If this filter is divided into the measured power spectrum ("deconvolution"), the result is the ideal one. Retransforming this, as shown in Figure 25, removes the blur and recovers an image with higher resolution and without the defects caused by the optics, motion, etc.

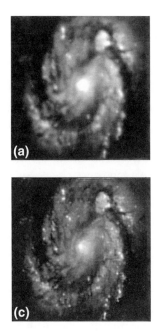

Figure 25 Image taken with the Hubble telescope: (a) original, with out-of-focus blur; (b) point spread function measured from a single star (enlarged to show pixels); (c) result from deconvolution of (a) with (b).

It isn't usually that simple, of course. The presence of random noise, particularly in the single star image, produces high frequency information in the blur function that, when divided into the Fourier transform of the measured image, produces large amounts of random noise that can completely mask the desired result. Some increase in background noise is evident in the example in Figure 25c, particularly near the edges. Many images with limited bit depth have too little precision to permit the mathematical accuracy needed for a good result. The method can be used with very low noise 8 bit images, but works much better with images from cameras with higher bit depths, as discussed in Chapter 1 (most astronomical cameras produce 12 to 16 bit images).

The other problem is measuring the blur function. In most cases there is no ideal, isolated, bright point source such as a star. In some microscope situations a small fluorescent bead can be used, and for some video camera work I have employed a tiny ("grain of wheat") bulb in a dark room to act as such a source, permitting an image of just the blur due to camera optics, electronic smear, etc., to be used to process actual images. The blur function must be acquired under identical circumstances to the actual pictures, and in most forensic photographic work this is not easily accomplished. An example is shown in Chapter 4, Figure 11.

There are other related deconvolution methods (e.g., Weiner filtering) that do not require the direct measurement of the blur function, but attempt to estimate it from characteristics of the image. These are iterative, extremely sensitive to the presence of any random noise in the image, and not usually very successful in improving the image resolution significantly. They are not dealt with here, since they are rather advanced methods that require someone with specific expertise in their use.

The other limitation of the deconvolution method is that it relies on the assumption that the blur applies equally to all points in the image. If portions of the image are out of focus because they are at different distances from the camera, or one portion of the image (perhaps a running person) has motion blur while the background does not, the deconvolution method cannot be properly applied. In theory, each fragment of the image might be deconvolved separately, but in practice obtaining or estimating the various point spread functions is not practical.

Measuring Image Resolution

Frequency space also offers a way to estimate the actual resolution of an image. Reference was made in Chapter 1 to the confusion that can arise between the number of stored pixels in an image, the number of individual sensors in the digital camera detector (or the number of samples taken by a framegrabber from an analog signal such as that from a video camera), the display pixels on the computer screen, and other numbers that may be stated as specifications that imply they are the resolution of the image. None of these is necessarily the actual image resolution, which is the size of the smallest feature or detail that can be distinguished from its surroundings or from another nearby feature.

Just as with photographic film, the image from a digital camera or computer can be enlarged on the screen to the point where the display pixel — the smallest element of brightness or color that the computer screen can separately control — is far smaller than the size of the features that can be distinguished. This is called empty magnification. If the digitized image is simply enlarged, the individual stored pixels can be easily detected on the screen, as shown in Figure 26. Sometimes the enlargement is performed using software that interpolates between the stored pixel values to fill in the space with new values. Interpolation can be done using a simple bilinear method that averages the values that surround the point, or using bicubic or other higher order methods that fit curves to additional pixels farther away. This can make the image look smoother and hide the edges of the discrete stored pixels, but it does not contribute anything new to the image and does not improve the resolution. Small features that were not resolved in the original image will not become visible in the interpolated image.

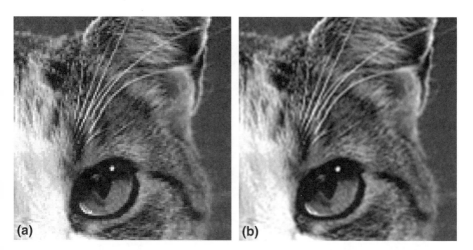

Figure 26 Enlargement of an image fragment to show individual stored pixels covering several display pixels: (a) expansion without interpolation; (b) expansion with bicubic interpolation.

One way to determine image resolution is to acquire an image of a test pattern with lines of known size and spacing, and see which ones can be distinguished in the stored image. It is not unusual for this value to be different in the horizontal and vertical directions (especially for cameras using video technology to acquire the image), so most test patterns provide sets of lines with both orientations. Figure 27 shows an example. It is important to note that the resolution of black and white lines gives a best-case estimate of the ability of the image to resolve small features. When the contrast is reduced, the detection of small features is worsened.

Changing lens settings or other optical parameters can affect resolution. Also, changes in lighting conditions can alter resolution, particularly in the case of dark images with a higher noise content that hides small detail. Consequently, the use of a test pattern to characterize the resolution of images acquired from a particular camera setup must be done under conditions that match those for the images, which is not always possible.

Resolution from the Fourier Transform

Another technique for resolution measurement uses the Fourier transform power spectrum of the image itself. Except for periodic structures that may be present in the image (e.g., in a photograph of a textile fabric, a fingerprint, etc.), which will produce spikes at the frequency corresponding to the spacing of the structure, the amplitude of the power spectrum falls off with increasing frequency. In most cases, a plot of the logarithm of the amplitude vs. the

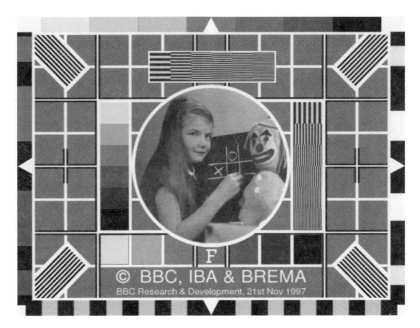

Figure 27 Typical video test pattern, which includes sets of horizontal and vertical lines to test spatial resolution and areas with uniform grey steps used to test tonal resolution. The pattern is also used to test for image distortion, and has areas of known colors to test for color rendition.

logarithm of the frequency will show a good fit to a straight line, but it is not necessary to use log frequency scales, and a simple plot of the intensity profile of the power spectrum, as shown in Figure 28, can be used. At a high frequency that corresponds to the resolution of the image the plot shows a definite change in slope, typically becoming essentially flat. The values at higher frequencies are nearly constant, and due to whatever level of random noise is present in the image.

Finding the change in slope, determining the frequency that it corresponds to, and expressing that value in terms of the pixels in the original image, provides a good estimate of the actual image resolution. This test can be performed in both the horizontal and vertical directions when appropriate. In the example of Figure 28, which is a single frame from a typical surveillance video, the plots show the profile of intensity vs. frequency for the horizontal and vertical directions, with superimposed lines showing the slope of power vs. frequency and the high frequency noise, with an arrow marking the intersection.

In the vertical direction, this intersection occurs at just about half the maximum frequency. The maximum frequency in a power spectrum plot corresponds to the ability to just resolve lines with a regular spacing of 2 pixels in the stored image. Half that frequency corresponds to twice the wavelength, and hence to the ability to result lines with a spacing of 4 pixels. The image resolution is thus 4 pixels in the vertical direction. The change in slope for the horizontal direction occurs at a somewhat higher frequency, about two-thirds of the maximum. By the same reasoning as for the vertical case, this corresponds to a resolution of about 3 pixels in the horizontal direction.

Examining the spatial domain image carefully confirms this estimate of resolution. The image has been formed with only one field from the video signal, as is often done with surveillance video. To generate the displayed image, each horizontal line has been repeated ("line doubling") to fill in the missing lines that would have come from the second interlaced field. Consequently, the resolution in the vertical direction should be about half of the maximum video resolution. Images can theoretically resolve line pairs with a pitch of 2 pixels (one pixel for a line, the next for the space between lines, the next for the next line, etc.). A measured resolution of 4 pixels is what would be expected for a line-doubled image produced from a single field.

Likewise, in the horizontal direction the smallest features that can be distinguished or the smallest spaces that can be identified between objects is about 3 pixels. The enlarged portion of the image shown in Figure 28d illustrates this. The narrowest vertical bright and dark structures (items on the store shelves) that can be discerned are about 3 pixels wide. A typical video image is digitized to form a stored image consisting of a 640 × 480 array of pixels, but in this case it is evident that the actual resolution is not this good. The vertical resolution is about 120 line pairs, corresponding to a useful image size of about 240 pixels, and the horizontal resolution is about 215 line pairs corresponding to a useful image size of about 430 pixels. Of course, an image formed with such display pixels would not have the correct 4:3 aspect ratio, but this is in agreement with the fact that the resolution is different in the horizontal and vertical directions.

As video images go, this is actually quite good. A video image recorded on videotape would be expected to lose some of its horizontal resolution. Most recorders specify no more than 320 pixels of resolution in the horizontal direction, and any dirt on the record/playback head or wear on the tape oxide can degrade this much farther.

Any image processing that degrades resolution is revealed in the power spectrum. In Figure 29 an image of a textile has been smoothed, as discussed in Chapter 2. The plot of the amplitude of the power spectrum as a function of frequency, averaged over all directions, is shown for the original and the smoothed images. The smoothing has obviously reduced the amplitude of high frequencies and shifted the transition point where the slope changes to a lower frequency, corresponding to poorer resolution.

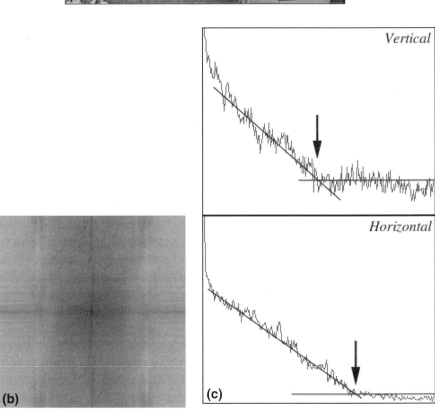

Figure 28 Measuring resolution on a surveillance video image: (a) original image; (b) Fourier transform power spectrum; (c) radial plots from (b) in the vertical and horizontal directions, showing the transition points discussed in the text; (d) enlarged portion of (a) to show individual stored pixels and the smallest elements of resolution present in the image.

Figure 28 (continued)

Tonal (Grey Scale) Resolution

It is also important to be able to ascertain the actual tonal resolution of a recorded image. Just because the image has been stored using brightness values from 0 to 255 (8 bits, which fits the organization of computer memory) does not mean that it has that much real tonal resolution. The criterion is the ability to distinguish two regions with a minimal brightness difference between them. Most video cameras deliver no more than 60 discernible grey levels, although many still cameras (because the image data is read out more slowly) can deliver more tonal resolution using the same detectors. Because the noise level of images generally increases as illumination levels drop, the values for images acquired in dim light, requiring higher amplification, are usually much poorer in tonal resolution.

The requirement is that the noise level in two different regions be less than the magnitude of their difference, which is the minimum condition needed to distinguish the regions. This can be evaluated using a series of grey scale steps on a test card, as shown in the example in Figure 27. A line profile across such steps shows the noise (Figure 30). It may be convenient to think of the problem as measuring the height of a stair step that has been covered with carpet. A small step covered with deep shag carpet may be obscured by the variation in the height of the texture of the carpet.

Rather than examining the line profile, it is often more convenient to use the image histogram. The noise in the uniform areas produces widths for the peaks that can be compared to their spacing. If two peaks cannot be distinguished, then the grey levels are too close together to be detected separately in the presence of the noise. Figure 31 shows histograms for the test

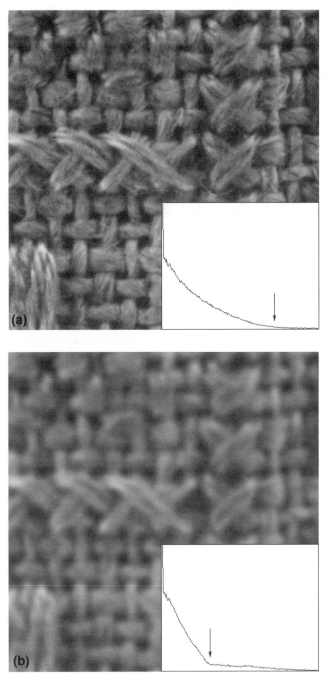

Figure 29 The effect of smoothing on image resolution: (a) original image, with plot of the power spectrum amplitude vs. frequency; (b) the same image and plot after application of Gaussian smoothing with a standard deviation of 4 pixels.

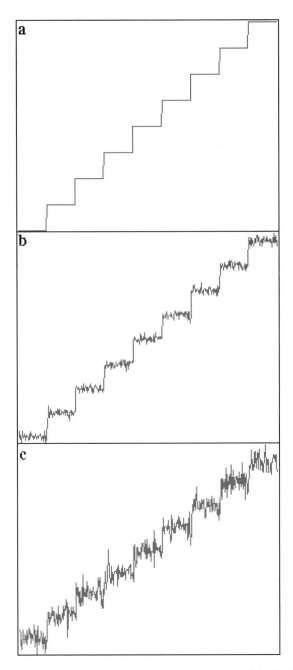

Figure 30 Intensity profile across a series of grey steps: (a) ideal case; (b) low noise — the steps are easily distinguished; (c) high noise — the steps are barely distinguished.

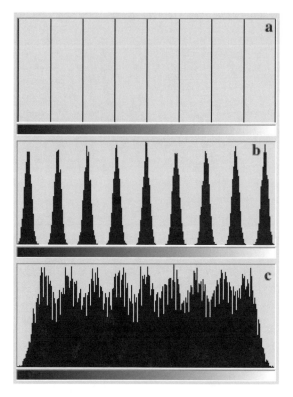

Figure 31 Histograms for the images of grey steps corresponding to Figure 30.

patterns from Figure 30. If no test pattern can be used to acquire data to represent and characterize the performance of an imaging system under the same conditions in which the actual images of interest were obtained, then it may be possible to use the histogram peak from any region in the image that should be uniform.

The width of that peak is a measure of the noise. The ratio of the total stored dynamic range of the image divided by that width is an estimate of the number of brightness values that can actually be distinguished. It is a somewhat optimistic estimate, since the peak widths may vary with brightness level. This is a problem particularly with CMOS chips, less so with CCD chips used in most high quality cameras. It is also optimistic because the steps in Figures 30 and 31 are shown as equally spaced, whereas some cameras (as discussed in Chapter 1) produce steps that are logarithmically spaced for equal intensity changes, which would reduce the distance between some steps.

Detecting Compression

Chapter 1 pointed out that a significant drawback of many digital cameras (particularly inexpensive consumer models) was the use of image compression to store images in less memory. This compression is lossy, and is accomplished by the elimination of some fine detail in the image, the shifting of detail and boundaries, and the alteration of colors. But it is widely used because the compression algorithms employed have been designed to retain those aspects of the original image that are useful to human viewers for the recognition of familiar objects in the pictures. Since images with forensic evidence are expected to retain as much of the detail as possible, in a scene that is likely to contain unexpected and unusual features, the use of lossy compression should be avoided.

But if you are given an image as a record of some evidence, how can you tell whether or not it has been previously subjected to lossy compression which might have altered or removed details, which in turn might be important? There are several different types of compression that can be used. A few of these, such as the LZW (Liv-Zempel-Welch) algorithm that can be used within the TIFF (tagged image file format) standard are loss-free, and simply look for repeating patterns in the image to encode them with no loss of detail. These methods rarely achieve significant reductions in memory storage requirements for real images, although they can often greatly compress computer-generated graphics.

Most image compression methods start by converting the image data from the stored RGB values to a more useful color space, one of the various HSI spaces, in which the hue, saturation, and intensity data are treated separately. The most widely used lossy technique is the current JPEG standard, which is based on a discrete cosine transform (DCT) much in principle like the Fourier transform mentioned above. The algorithm divides the image up into small (usually 8 × 8 pixel) segments, applies the DCT to each, and quantizes the coefficients into discrete value steps (the greater the compression, the fewer steps available and hence the poorer the precision with which the values are represented). This process discards terms with very small amplitudes and uses few digits to represent the ones that remain. The algorithm then encodes the terms by saving only the difference from one region to the next, and routinely achieves compression by a factor from 10 to 100 times.

There are also other competing technologies. Wavelet compression, which will become part of the JPEG standard within the next few years, replaces the sinusoids used in the DCT and FFT with a different set of basis functions, called wavelets. There are several such functions, each with particular advantages for ease of computation and retention of detail. The advantage of wavelets over sinusoids is that they can localize detail better and generate fewer "ringing" artefacts at the edges of images. The power spectrum from a Fourier transform often shows a cross pattern resulting from the large number of frequencies needed to represent the difference in image content from the top edge to the bottom, and from the left edge to the right. Wavelets do not have this problem, and generally deal better with the edges of images.

Fractal compression examines small blocks of data in the image and finds larger blocks that have the same local pattern of contrast. The location of the larger block and the rotation of the pattern are stored. Reconstructing an image from this set of relative offsets simply duplicates the patterns. The method often achieves compression ratios greater than 100 times, and can apparently enlarge images by continuing the process of copying patterns, reducing them in size, and inserting them in the reconstructed image as required. Of course, the patterns contain detail that makes these expanded images appear to have information all the way down to the pixel level, so the Fourier transform power spectrum will not show a limit to the image resolution. However, the detail that is inserted is not real, just something borrowed from elsewhere. The manufacture and insertion of apparently real detail is a serious concern for forensic applications in particular.

The alteration of images by these lossy compression methods can be easily demonstrated by simply subtracting the original image from the reconstructed compressed image. The differences (see the example in Figure 32) are typically in the fine detail and location of edges, which people generally ignore for purposes of recognizing familiar objects. There is also, in most cases, a pattern of vertical and horizontal lines that correspond to the size of the block used for the compression method. Almost all compression techniques (except for a few types of wavelet transform) break the image up into blocks as part of the algorithm. Transform methods do this because it is so much faster to process small blocks compared to the entire image. Also, the block-by-block compression facilitates the transmission and reconstruction of images — the principal reason for the development of the technology — block by block, instead of waiting for the entire image. The fractal method uses the blocks to search for larger ones with similar contrast patterns (Figure 33).

If an image is received without information on its prior history, or if the use of a lossy compression technique is suspected, but there is no way to compare it to the original, the presence of blocky artefacts in the image is a revealing clue. Applying a high pass filter such as a Laplacian to the image will usually reveal the block boundaries, and plotting the intensity profiles in the horizontal and vertical directions will show regularly spaced peaks. There are some sources of periodic noise such as interlace jitter in video, and

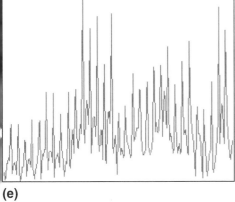

Figure 32 JPEG compression of an image: (a) original; (b) reconstructed from JPEG compression, which reduced the file size by a factor of 5 (more compression increases the magnitude of the effects shown here); (c) difference between (a) and (b); (d) application of a Laplacian sharpening operator to (b); (e) intensity profile of (d), showing regularly spaced spikes (a spacing of 8 pixels, corresponding to the JPEG block size).

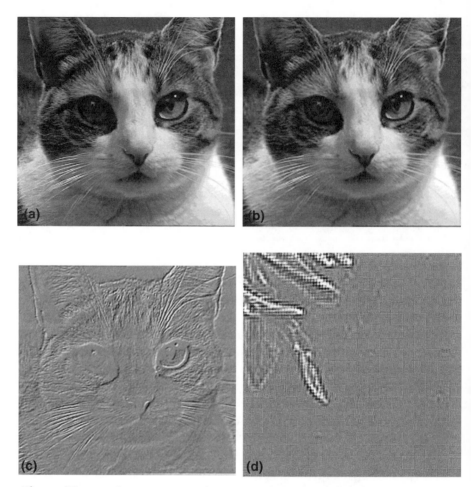

Figure 33 Fractal compression of an image: (a) original; (b) reconstructed from fractal compression (17:1 compression); (c) difference between original and fractal compression result (details have been created and shifted, sizes have been changed, and in a color image the colors will be altered); (d) example (enlarged) of blocky artefacts produced by application of the Laplacian sharpening operator to an image that has been compressed.

high frequency clock noise in analog to digital converters, that can produce such peaks in one direction or the other, but not both.

Wavelet compression can be the most difficult to detect, although its effects on the image are easily evident visually when comparison to the original image is possible (Figure 34). The color information is degraded more than intensity data, so that colors bleed across edges. Also, detail is not lost uniformly across the image, but tends to be preserved in some places and practically erased in others. If an image is suspected of having been subjected to wavelet compression, the most reliable indicator is the presence of short horizontal and vertical black and white steps along the edges of features after application of a Laplacian filter.

In a few cases where these methods for detecting compression are ambiguous, or where it is important to know just which compression method has been employed, a more definitive test is to compress the image in question using all of the various compression techniques (for wavelet compression, this means using each of the popular wavelet basis functions). In the compressed data produced by the same compression technique used originally, there will be terms that are missing or quantized into a few discrete steps. This cannot happen by accident with an uncompressed image, and is a sure sign of the previously applied compression.

There are a few applications in which compression is accepted, having been tested to verify that it does not remove needed detail. The most important example is fingerprint images, as used in the automated fingerprint identification system (AFIS). The spacing of friction ridges on human fingerprints is quite uniform. Because the important detail for identification of friction ridge patterns is confined to a narrow set of frequencies, and the actual amount of data needed to locate the key points in the pattern for matching and identification is small, compression using a specific algorithm based on wavelets has become the standard for storing these images.

Detecting Forgeries

One of the enduring, although often vastly overstated, concerns about digital imaging is the belief that forging such images is so easy that they cannot be trusted. When the provenance of an image is not known from independent records, it can be important to be able to detect any editing that may have been done. Copying and pasting portions of one image onto another could in principal be used either to insert new information or to hide existing details, or to reposition them within the image.

This is not unique to digital imaging, of course. Retouching or creating composites in the darkroom can accomplish all of the same things, and has been employed for such purposes for as long as photography has been in use. But because it is difficult to do and requires special skills, the darkroom exercise is costly and usually is limited to special cases. Digital imaging and computer-based processing make it easier to learn the necessary skills, although they still require time, care, and considerable practice to master.

As evidence of the changes that computer-based imaging has produced, consider the use of special effects based on image editing in the movies. Removal of power lines or safety harnesses from scenes, superimposition of actors working in front of a blue screen onto scenes shot elsewhere, creation

Figure 34 Wavelet compression of an image: (a) original; (b) reconstructed after 38:1 wavelet compression; (c) difference between (a) and (b); (d) application of a Laplacian sharpening operator to (b) (enlarged to show detail). See color plates 25 to 28.

of artificial or animated characters that interact with human actors, creation of scenery or objects rather than constructing physical sets, and other similar image editing tasks, can be found in movies dating from the very beginning of the art. In most cases these were easy for audiences to recognize and used only in a few selected instances before the advent of digital technology. Now entire movies are made that way, and the use of these techniques has become both routine and extremely difficult to detect.

The history of fakery in photographs is nearly as old as photography itself, and has been a source of concern and controversy just as long. It even predates photography, since artists who painted scenes and portraits have always altered them for various aesthetic or other reasons. The same techniques of photomontage (combining portions of different images) that let Tom Hanks as Forrest Gump meet President Kennedy, or Fred Astaire dance

Figure 34 (continued)

in a modern vacuum cleaner commercial, were used during the Civil War to insert additional carnage into battle scenes or place Generals Lee and Grant side by side in a portrait.

As detailed in *Photo Fakery* (1999, Brassey's Publishers, Dulles, Virginia) by Dino Brugioni (a former CIA official whose task was to detect forgeries in photographs), most "serious" news organizations claim that they never edit an image to alter the "journalistic content." But that doesn't prevent removing details that they don't consider important, or improving the appearance of celebrities, or cropping images to eliminate portions and perhaps alter the impression they create, or printing photographs darker to make a suspect appear more sinister (as actually happened in the O. J. Simpson case). Whether these altered details will become important later is, of course, impossible to predict. And beyond straight news stories, our acceptance of the manipulation of images in advertisements, the enhancement of glamour photos, and the creation of montages with celebrities on supermarket tabloids, makes many people suspicious of photographs in general. We've seen the often rather amateurish efforts of former communist countries to edit out people from news pictures after they have fallen from political power, and expect western sources to do the same thing, but perhaps with more skill.

With the investment of enough time and skill, a forgery that cannot be detected can be produced by either film or digital technology. Even with the use of digital image processing in the computer, however, this is not something that is easy to do or to hide. It is not possible to give a complete listing of the signs to look for as evidence, but the list below is based on the four principal techniques for faking images: the removal of detail; the insertion of detail; combining portions of different images; and mislabeling of pictures.

Presumably the crime scene photographer will not be trying to perform any of these actions intentionally, i.e. by removing a feature from an image and replacing it with background, or using copy-and-paste to insert part of one image into another. But it requires some care to prevent unintentional actions from creating the same effects. A detail recorded in an image may not be visible depending on how the image is displayed or printed (e.g., a weapon may lie in a dark shadowed area or a stain may have little contrast with its surroundings). Whether the image is manipulated in the computer or the darkroom, the final viewed print can rarely show all of the captured detail and if something important is not visible that omission may be critical. Similarly the selection of the image field of view is intended to contain all important parts of the scene, but if something is inadvertently omitted then it will not be observed in the final photograph. Careful record keeping should eliminate errors in the description of pictures, but ultimately this relies on the records, testimony, and credibility of the photographer and not on the pictures themselves.

There have been criminal cases in which overt photo manipulation has been used, often by police and prosecutors. Crime scene photos released to the press may omit some crucial details which are intentionally withheld as a means to elicit revealing statements from suspects. In some cases, altered or even staged pictures of "victims" who managed to avoid being killed have been used to trap those who tried to arrange murders. But these pictures are not normally involved in trial situations. If photographs (digital or otherwise) are introduced by the parties to a civil or criminal trial, and the supporting testimony is inadequate to assure the jury of their fidelity, then it may be necessary to try to detect within the images themselves the evidence of any tampering, intentional or otherwise. This can be extremely difficult, and there is no guarantee that all forgeries can be uncovered. The principal role of the computer and digital imagery in this regard is that the process of image alteration has been made more convenient and rapid, and perhaps accessible to a broader segment of the population, than was the case with darkroom-based methods in traditional photography. The basic techniques and the primary errors that lead to detection have not changed.

Some of the most common errors that permit detection of forgeries in images are:

1. Features that are inserted into an image may be incorrect in size, may be oriented inconsistent with their surroundings (e.g., a weapon not lying properly on the supporting surface or a blade not connected smoothly to a handle), or may have different perspective from the surroundings. When inserting writing or other lines, it is often difficult to match colors exactly. Also, for writing, both the width of lines and the variation of that width as a function of local line curvature may be diagnostic.

2. The brightness, color, or lighting on the object may not match the rest of the image. The most common problem is that the direction of lighting may be different. This also extends to shadows, which may point in the wrong direction, have inconsistent dimensions, or may not fall properly onto other portions of the image. If one part of the image has been copied from another image or from another part of the same image and pasted into a scene, it is difficult to match the brightness and contrast of features and the shadows they cast at the same time. Highly directional lighting and strong shadows can be matched with enough effort, but the effects of diffuse ambient lighting which is reflected from other objects in the scene (especially colored ones) are much more difficult to match exactly. The presence of mismatched shadows in different parts of the scene is usually taken as evidence of tampering. Color analysis in general is a powerful tool to detect forgeries within an image. Matching the colors of objects that should be the same, particularly when they have different illumination, requires very careful and painstaking adjustment.

3. The focus of the inserted object may be inconsistent with other features in the image at the same distance from the camera. This is particularly difficult to fake when the changes are made in the scene covering a range of distances so that the focus varies. In some cases, access to the camera used to take the pictures may be useful to determine whether the optics are consistent with the recorded image (e.g., depth of field, distortion of straight lines, or vignetting or darkening of image corners).

4. The texture of the object may not match other features in the image that have similar illumination. In conventional film-based photography this is most often studied by examining the size and shape of the grain structure of the negative, which varies with type of film, exposure, and

development and is very hard to match from one negative to another. In digital images the pixels are much larger than this grain size, but it is still possible to examine the speckle or brightness variation in the pixels within the feature and within other similar regions of the image. Differences in the speckle can be measured using statistical parameters such as the local variance or entropy of the pixels, or by comparing the breadth of peaks in the histogram. The particular illumination, optics, and camera characteristics — especially the type of detector used — produce a noise pattern in the image (both random noise and any periodic noise that may be present) that is not easy to match when pasting in part of another image, or when painting out a feature that is being removed. Pasting in images of objects often requires smoothing along the edges to blend the image into the scene, and this lack of noise at edges can be detected. In some cases, it may be useful to plot the noise level as a function of brightness, and compare suspect regions to this trend. Different cameras and illumination levels typically produce different amounts of random noise. Periodic noise can be a signature for the system used, as it generally results from specific defects in or characteristics of the hardware (e.g., the type of color filter used on the CCD chip). It is often most easily detected using the Fourier transform. A crude way to attempt to hide texture and noise differences due to editing is to add more random speckle to the entire image afterwards, to try to overwhelm the differences introduced by editing, but this is usually simple Gaussian additive noise and does not vary with brightness as it would in a normal image. The presence of such added noise is itself often taken as a *prima facie* sign of manipulation.

5. Technical or scientific inconsistencies may be present in the image. For example, an automobile speeding down a dusty road should leave a trail of dust, or a boat moving in the water should leave a wake. The damage to a crashed vehicle must be consistent with the reported speed at which it was traveling, etc.

6. In a few instances, a forgery may be detected because another picture is found somewhere that contains the features that have been copied into the forged image. In many doctored "news" images, stock pictures are used to insert backgrounds, vehicles, or other individuals, and the same images appear in more than one photo. When a detail is removed from an image, the background used to replace it may be copied from elsewhere in the image, which can be detected by a repetition in the pixel values.

7. When more than one photo is available, either a series of images from a video camera or simply multiple images taken at the same scene from different viewpoints, photogrammetry can be used to determine

whether the placement and dimensions of objects are consistent in the various images.

8. For outdoor images, the sun angle determined from shadows can be used to determine the time when the photograph was taken, often to the exact day and hour. It is necessary to know the location of the photo and direction of north. Conversely, if the time of a photograph is known (which many digital cameras record in the file), the latitude and longitude can be determined.

9. When portions of two images are merged, the background details are often inconsistent. Efforts to make the foreground objects match in terms of orientation, lighting, brightness, etc., commonly overlook differences in the backgrounds, which may be simply blurred and blended together. Details in the background such as walls, floor coverings, wildlife, etc., may not match.

10. When an object is removed from an image, details such as shadows or reflections may remain to betray its presence.

Figure 35 shows an example of an altered image with several noticeable defects:

1. Removal of an object has left the shadows that it cast onto other portions of the image.

2. The object was removed by copying and pasting a region of the background from elsewhere in the image. This is evident because the grain pattern (and the seam in the table) exactly matches that in another region.

3. An object has been added to the image. The coffee cup has perspective that does not match its location on the tabletop, and the angle of the light source is slightly different from the rest of the image. The cup does not cast a shadow onto the table. Also, the details on the cup are much sharper than other nearby parts of the image; this could arise either because the image was captured at larger scale and reduced, or was taken in sharper focus.

It is very difficult and time consuming to do major editing without leaving some traces. Adding a person to a group scene, or removing one, is difficult. Repositioning portions of features or limbs of people, and making more subtle changes in image evidence can be even harder. But it must be emphasized that with enough care a successful forgery can be made. Some relatively simple modifications (e.g., repairing or adding a hole in an image of clothing) can pass any inspection method, just as can be done with film images. Of course, this is the reason that it is vital to have independent data

Figure 35 An altered image with visible defects as discussed in the text.

and reliable testimony on the history of each image and to use typical evidence chain-of-control for the original image files.

Maintaining Records

The controversy over the use of image enhancement techniques is a meaningless one when the tools are properly applied and explained. No image, whether processed or not, digital or film, stands by itself. It is the integrity of the witness presenting the image that must be demonstrated. Physical items themselves are not "evidence." The testimony of a witness is evidence, and the image is an "exhibit" to that testimony. The methods described in this chapter and the preceding one are well known in the field, widely accepted, taught in most textbooks and courses, and well able to pass *Frye* and *Daubert* tests for admissibility. The only remaining problem is maintaining adequate records to show what has been done to the image, and to convince the jury that it was properly done.

The first part of the task is easy enough, although it can be time consuming and requires the care associated with any examination or processing of evidence. It is necessary to keep a complete log of all steps carried out with each image. In many systems, this can be done automatically by having the computer construct the log. This file can be recorded along with the derived copies of the image, on a permanent archive such as a CD-R, particularly one with permanent serial number coding. In other situations, if the software does not keep a sufficiently detailed record, it is necessary to do so manually, preferably in a bound lab notebook with each page signed and dated. It may still be necessary to justify the use of the techniques, but at least there is no question about what was done.

When presenting the results of image enhancement, it is usually not very wise for the expert to simply show the results and state that the work was done because he or she (the expert) knew it was proper. Juries may not be expert in the science of image processing, but everyone thinks they understand images and wants to feel a certain level of comfort about what was done. And no one likes to be told they aren't smart enough to understand the processes used or the reasons for their selection.

It is generally very helpful to prepare a series of step-by-step examples to show the process of enhancement, and to compare the results at each step to the original to show that no details have been introduced or eliminated, but rather that the enhancement methods simply make details that are really present in the original image more visually evident. This can be done either as a series of discrete images representing the various steps, or as a continuous movie showing the evolution of the image and the improvement in visibility of the information of interest. Descriptions in lay language of the processes involved, their effects on the image, and the general types of applications that they are used for, are invaluable for educating and informing the jury, and demonstrating that the expert witness' opinions have a sound basis and should be trusted by the jury.

Identification

4

Many facilities, including banks and convenience stores, are "protected" by having surveillance video cameras and recorders installed. Sometimes, as in most commercial establishments, these are prominently visible as a deterrent to would-be thieves. In other cases, they may be well hidden with only a peephole lens mounted somewhere high on a wall or concealed in decorations. In all cases, there is the expectation that the recorded images will be useful for the identification of persons, objects, and actions recorded on the tape. This goal of identification actually has several levels. In cases of robberies, the images are often shown on local TV news in hopes that someone will recognize the perpetrator and call police with the information. Once a suspect has been arrested, the videotape images may convince him or her to admit guilt. Failing that, perhaps the videotape images will convince a jury that the correct person has been apprehended and secure a conviction. In other applications, surveillance video may be used to capture license plate images (e.g., to issue speeding tickets or perform security checks at protected installations). This latter application will become significantly easier when all states change the paints used to provide a clear image of the numbers in the near infrared, which can be detected well by solid-state cameras. At present, some of the design graphics used by states obscure the numbers.

The Imaging Chain

To understand the limitations of these images, it is necessary to go through the entire imaging process and consider the limitations of each component. Figure 1 shows a schematic diagram of some of the elements in a surveillance video system. Many of the difficulties start right with the very first element: the location of the camera and the lighting in the room. It is axiomatic that if the camera is placed where it cannot see the person, or cannot capture an

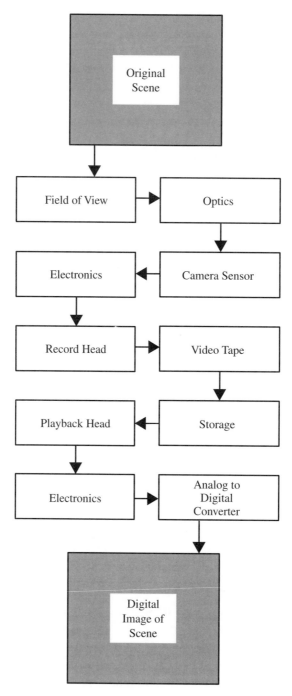

Figure 1 Block diagram of a typical surveillance video system.

Figure 2 Surveillance video from a convenience store, showing problems discussed in the text.

image of the person's face, then the images will be useless. But in many convenience stores, for example, the camera is likely to be blocked by stacks of goods for sale, or signs hanging from the ceiling, etc. Figure 2 shows a representative example. The person at the counter cannot possibly be recognized from this video image because his face is behind a sign posted at the checkout location, and the person near the door, who might be an accomplice, is hidden behind a pile of boxes.

I have had police approach me with images of this type and ask if by image processing I can remove the sign or box to reveal the faces of the perpetrators. These people don't understand what image processing can and can't do, and have probably watched too many movies like "Rising Sun" in which the technology is vastly misrepresented in its capabilities, the quality of the images, and the ease and speed with which results are achieved. They probably also believe that satellite images can zoom in to read newspapers.

Also evident in Figure 2 is a second problem with camera location. The view is from the back of the store towards the front, so that light coming through the windows dominates the scene and faces are generally seen only as silhouettes, with little contrast. Even in nighttime scenes when there is no external sunlight to deal with, the camera placement often includes ceiling lights which produce a similar contrast problem, as shown in Figure 3. Of

Figure 3 Nighttime surveillance video with ceiling lights.

course, the lack of visible contrast does not mean that enhancement by manipulation of the grey scale values (contrast expansion and brightness adjustment) may not be able to recover some information, but it certainly makes it harder. As a side note, the problem is worse for persons with dark skin pigmentation, where there is often little contrast to begin with.

In many cases, the camera location in a store is set up to watch the cash register, more to discourage employee pilfering than in a serious attempt to identify thieves. Also, the optics may be set up with an aperture that is a compromise between daytime and nighttime conditions, so that the daytime images are too bright and the nighttime ones are too dark.

The optics can have other problems as well. The lenses used in these situations are usually pretty wide angle, inexpensive ones that introduce distortions and may not have enough depth of field (assuming they are properly focused at all) to focus on the entire scene. Low or no maintenance can result in lenses that are knocked out of correct focus or dirty, which also degrades the images. A wide angle lens may provide coverage of the entire store, but the faces will be too small to record any useful detail (see the discussion on resolution, in Chapter 3 and below). A longer focal length would make the faces larger in the image, but only if they happen to be in the much smaller field of view (Figure 4), and in focus (the longer the focal length, the shallower the focal depth).

Figure 4 Scenes of a convenience store using lenses with varying focal lengths and fields of view.

Surveillance Video Cameras

Next comes the video camera. A grey-scale camera used for surveillance imaging can be purchased for a few hundred dollars, and it is unusual for anyone to invest more in a camera of higher quality or with color capability. The majority of these cameras use CCD chips of rather limited performance specifications, and it is likely that even less expensive CMOS devices will capture a large share of this market in the near future (with even poorer performance, especially as regards noise and low light sensitivity). Some surveillance cameras use older vidicon technology with an evacuated glass tube and a scanning electron beam. In principle, these types of tubes can provide high spatial resolution and good tonal resolution, but they are strongly affected by stray electromagnetic fields, vibration, and degradation with age.

The CCD chip used in a high quality digital camera may have several million discrete diodes that detect light to form a high spatial resolution image, and a slow readout to preserve 8 bits (256 brightness levels) of tonal resolution. For surveillance cameras, the emphasis on low cost encourages using the smallest possible detector (1/4 inch or even 0.2 inch diagonal measurement, rather than the 1/2 or 1 inch chips used in more expensive cameras). This means that the number of diodes and the size of the individual diodes must be limited. The number of rows of scan lines in the image is fixed by the video specification (480 for standard US-style video), but there may be fewer rows of detectors in the camera (for example, half as many rows might be used, with one row used to generate the even numbered scan lines and the average of two adjacent rows used for the odd numbered scan lines). Likewise, the number of diodes on each row is not specified. There are some cameras with only 400 detectors along each row, whereas a high quality consumer camcorder may have 750 to 800, and a professional camera even more.

Using smaller individual detectors captures less light (and since the spacing between detectors is not reduced, less of the total light entering the camera is detected). This reduces the camera's sensitivity for dark scenes, requires more electronic amplification, and increases the noise in the images. The use of inexpensive lenses with small apertures also reduces the amount of light available.

In earlier chapters, the distinction has been made between the number of detectors across the width of the image (sometimes called camera pixels), the number of samples taken by the framegrabber or analog-to-digital converter, which usually corresponds to the number of stored pixels in computer memory, and the number of display pixels on the computer screen. This is very important. It is the lowest of these values that sets an upper limit on the number of resolution elements that may be present in the image. There are other sources of degradation, including electronic noise, dirty video recorder heads, worn tape, random or periodic noise, and improper image processing, that can reduce the resolution further, but it can't be better than the weakest link in the chain shown in Figure 1.

The specification for broadcast video is based on the bandwidth assigned to each channel, and limits the resolution of the received image to a maximum of about 330 resolution elements across the width of the video screen. Most displayed images are somewhat poorer than this because there is overscan that extends off the sides of the display, and a figure of 300 resolution elements is typical of actual performance. The vertical resolution is 480 lines of image, but again overscan typically reduces the actual display to about 400. Notice that the word pixel has not been used here — the video signal is analog and there is no defined pixel on the video display. These values correspond to NTSC video (NTSC is the National Television Standards Committee), as used in the US, Canada, and some other countries. European video (PAL in most countries except France, which uses its own system) has more lines and slightly more resolution elements per line, but presents fewer images per second (25 instead of 30).

The clever trick that allows broadcast video to show motion without objectionable flicker with only 30 frames per second is interlacing two fields to make up each frame. In each sixtieth of a second, only half of the lines in the image are scanned; in one such field the odd lines are scanned (1, 3, 5, ...) followed in the next field by the even ones (2, 4, 6, ...). This is fine for fooling the eye, but it creates many problems for image acquisition. When a video camera is used directly connected to an analog-to-digital converter (also called a framegrabber) in the computer, the interlaced fields are combined to make a frame. If the two fields do not line up properly, either because of electronic jitter or motion in the image, it degrades the frame image significantly.

Recording and Playback Problems

Most video recorders (which are designed for the consumer marketplace to record and playback typical video programs) record both fields sequentially and play them back in the same way. These recorders have no more resolution than that needed to record the broadcast signal, so the highest horizontal resolution they offer is 330 samples (resolution elements) per line.

Many years ago, broadcast television introduced color. The broadcast channels were already established and their bandwidth was not increased. Also, there was a strong desire to make the existing black and white sets capable of receiving the color image. The solution was to further reduce the bandwidth (the luminance data is broadcast in the bandwidth below 3.58 MHz, with the higher band reserved for the chrominance data) and hence the resolution available for the intensity information, to no more than 220 samples across the width of each line, and to squeeze the color information in with about half that number of samples across the width of each line. Because human vision is rather tolerant of color smearing, the use of poorer resolution for the color signal was considered acceptable.

The recorders used for surveillance work do not generally record the full video signal, poor as it is. Standard videotape cassettes can hold no more than about 2 hours of continuous video (low speed recording can squeeze up to 6 hours on a cassette, with correspondingly less resolution), but surveillance camera tapes are typically changed daily. Instead of recording everything, they usually record only one field, not the full frame, which reduces the number of video lines by half. Also, they usually capture one image every few seconds, rather than recording continuous motion. Recording 2.5 fields per second will fill a one hour videotape cassette in 24 hours, and many recorders are not able to record the intermittent images with full density on the tape.

The immediate consequence of this strategy is to reduce the maximum resolution of the video images to about 220 samples across the width of the line if the camera produces the full video signal, including color information (even if color is not being recorded, most cameras put out the full RS170a color signal since it allows them to use standard electronic components), or 330 samples if monochrome signals are produced by the camera, and about 200 lines per image. Note that these resolution elements will limit the ability to visualize small details in the image, as discussed in earlier chapters. This point will be expanded later in this chapter.

While the camera and recorder may be capable of capturing this amount of resolution on tape under optimum conditions, the typical surveillance camera setup is far from optimum. The wiring from camera to recorder may be long and may use cables with incorrect impedance, which will reduce the bandwidth and hence the resolution, and the camera electronics may be misadjusted in terms of gain and brightness (sometimes this is done to compensate for incorrect aperture settings on the lens).

The greatest loss of resolution comes from the actual recording process itself. Dirt builds up on the record head, which moves over the oxide-coated tape at an angle during recording and playback and collects loose oxide particles (and any other dirt on the tape) in the gap. This build-up of dirt forces the head away from the tape, reducing the high frequencies in the recording that are responsible for the maximum resolution. It is the high frequencies that represent edges and fine detail in the image. In addition, wear on the tapes, which are typically re-used many, many times, removes oxide so that the full magnetic field strength is not recorded, and this also reduces high frequencies more than lower ones.

The playback process has all of the same problems. Dirt separates the tape from the head, and produces a lower signal strength which must be amplified more, introducing more noise and loss of resolution. The fact that only one interlace field was recorded is compensated for in the electronics by "line doubling," which means that each scanned line is repeated to keep the full height of the video image without missing lines. Of course, this does not fill in any new information. The consequence of these factors is that actual surveillance tapes rarely provide even the rather poor resolution of which the technology is capable. Measurements on actual images rarely show more than about 200 resolution elements across the width of the image.

Pixels and Resolution Elements

The multiple meanings of the word "pixel" have been discussed before. In this instance, the confusion arises once the analog image is digitized into a computer for analysis, enhancement, presentation, and possibly printing hard copies. The most widely used video framegrabbers produce an array of 640 × 480 stored pixel values. The aspect ratio (4:3) of this array of square pixels is correct for NTSC video, and the size corresponds to standard VGA computer screens, so these are convenient dimensions. But these stored pixels are much smaller than the resolution elements in the analog signal. Even without further expansion on the computer screen, the stored image represents empty magnification of the actual image data. As shown in the example in Chapter 3, Figure 28, the resolution elements may be several display pixels high and many pixels wide.

Of course, not all of the pixels within the resolution element have identical values. This is in part a measure of the noise in the system and in part a continuous smearing or interpolation of the analog signal that makes up the image. It is difficult and quite time-consuming to measure the noise level and blurring of each component of the system shown in Figure 1. Using test pattern generators, many of the transmission, recording, and digitizing steps can be quantified. By using a test pattern chart in front of the camera, the additional contributions to system performance limitations caused by the camera can be determined. It is important in such tests to use lighting levels and exposures similar to those of the scenes of importance. Usually the results will be far worse than the specifications of the original new equipment. But it does little good to perform these measurements, since the resolution of the final image is what matters, not the performance of the system or its components. It is usually more useful to directly estimate the image resolution as discussed previously.

Describing the digitized image in terms of its size in pixels is likely to be misleading, as compared to the resolution limits imposed by the imaging system. I prefer to describe these as resolution elements to avoid confusion with the word pixel. If the image has (for example) an actual resolution of about 200 points (or 100 line pairs) in the horizontal direction, then that is the criterion that can be used to judge the smallest features that can be confidently discerned and used for recognition.

Figure 5 shows a fragment from the image in Figure 4c, after adjustment of brightness and contrast but no other enhancement or processing. The person's ear in this image is 12 stored image pixels high and 6 wide (it has actually been further enlarged on the screen image, to twice that size). If there are 200 resolution elements across a 640 pixel stored image, then by simple ratioing the image of the ear consists of about 10 resolution elements (5 high by 2 wide). Visual examination of the image will confirm this: there is a single dark element that corresponds to the center of the ear, a few bright ones that make up the ridge, and so forth.

Figure 5 Detail from Figure 4c.

We know it is an ear because of the position on the head, and because we have seen a lot of ears we impose our memories of what ears should look like on the image. But in fact there is very little information there to match the image with the ear of any particular suspect. Only a person with a very unusual ear, major mutilation, etc., would be rejected as a possible match.

It is important to reiterate here the distinction between resolution elements and the number of pixels in the image. In one trial situation the prosecution used the example shown in Figure 6. The enlarged image of the perpetrator's head uses about 20×20 display pixels for the ear. However, this does nothing to improve the actual image resolution (about 3×4 resolution elements for the ear) or the ability to identify the subject.

Furthermore, because of the noise in the image which superimposes a speckle pattern on the displayed resolution elements, successive images of this same ear will be slightly different. Which one is correct? Can we choose the one that matches (or does not match) the suspect and use that as evidence? No. Clearly the evidence represented by the image of the ear is ambiguous in terms of identification of any particular individual. Similar situations apply to each of the individual facial features in the video image (mouth, nose, eyes, chin, etc.). With the wide angle lens picture (Figure 4a) the situation would be even worse.

Figure 6 Enlargement of a portion of a surveillance image.

Noise and Tonal Resolution

Television signals originating from expensive studio cameras routinely produce images with 100 or more distinct brightness levels. This amount of tonal resolution is more than people can actually discern in the displayed video image or in a photographic print. Even relatively inexpensive cameras used in surveillance work are capable of producing signals that can identify 50 or more grey levels when used in a well-illuminated environment, and this is still more than needed to satisfy the needs of human vision.

But as the scene illumination decreases and camera gain increases, the noise increases as well. Part of this is the statistical fluctuations in the electron signals output from the individual detectors, and part of it is due to the electronic and thermally generated fluctuations produced in the transfer and amplification of the signal. When the scene contains areas of pure white or black, such as the bright background from windows in a surveillance image, this may affect the amplifier gain (because most cameras use automatic gain circuits to adjust the overall image brightness) to make the noise situation worse, and to compress the useful portion of the signal so that the ratio of signal to noise decreases further.

After the camera, each component in the system adds more noise. The recording and playback process is particularly guilty of this; a dirty record head or worn tape that has lost oxide particles cannot record subtle variations in brightness accurately. Incompletely erased prior images may contaminate the current ones. Finally, the analog-to-digital converter or framegrabber

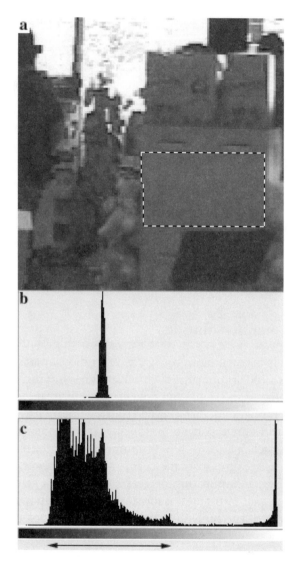

Figure 7 Measurement of tonal resolution: (a) selection of a region in Figure 2 that should be uniform in brightness; (b) histogram of that region; (c) histogram of the entire image. The arrows mark the range of brightness values from the interior portion of the scene; the peak at the white end of the histogram corresponds to the backlighting from the windows.

may also introduce noise. Some of this noise is random, while some may be periodic noise due to the proximity of high frequency signals within the computer, which radiates onto the capture board. The effect of the noise is to reduce the ability to distinguish small steps in brightness, thus reducing the tonal resolution of the image.

In Figure 7, a selected area corresponding to the side of a box that should ideally be uniform in brightness is shown with its histogram. The peak width at half height is about 7 grey levels, which suggests that as many as 36 grey levels (256/7) might be resolved in the image. The actual number is somewhat less, because the peak width typically varies with illumination level and the output from the camera is linear, whereas human vision responds logarithmically to brightness, but even so there might be more than 20 to 24 resolved grey levels which should be enough to satisfy the needs of observation.

However, in this case, the presence of a bright background from the window has reduced the tonal range available for the useful image information. As shown in Figure 7c, only about half of the total range of grey levels is used by the interior scene. Hence the actual tonal resolution is also reduced by about half, leaving only about 10 to 12 useful grey levels of tonal resolution to form the important elements of the scene. Adjusting the brightness and contrast of this stored image can redistribute the grey levels, but it cannot increase the tonal resolution.

Other Shortcomings

In addition to the problems of spatial and tonal resolution, surveillance video has other important shortcomings for identification purposes: lack of color and motion information, and lack of sound recording. Color cameras are more expensive than black and white, which is the principal reason they are not used in most surveillance applications. Even if they were, calibrating the color information in a scene would be very difficult, because of changing scene illumination and camera gain. The color information in NTSC video is not very good under the best of circumstances (but many engineers joke that it really means "Never The Same Color" because the reproducibility and quality are so poor), and the resolution is at best half that of the brightness portion of the signal.

But color can be a very important piece of information for identification purposes. More important than skin color and tone, which can often be judged from grey scale images and whose subtleties aren't reproduced all that well in video anyway, is the identification of clothing color or other distinguishing marks, or the color of automobiles (sometimes the vehicle can be seen through the window in convenience store burglaries, and of course there are other surveillance situations that directly involve vehicles). It has been pointed out before that human vision does not distinguish brightness levels very well, but we are quite good at noticing color variations. We don't "measure" color visually, don't all use the same words to identify colors, and one person's description of a color may not agree with another's. But knowing that the perpetrator wore a "greenish" jacket — whether it was actually green, olive, teal, turquoise, etc. — is much more useful than just the grey scale picture.

Less obvious, but also missing from typical surveillance videos that capture one field every few seconds, is any information about motion. Some people have characteristic walks, arm motion, etc., that can help in recognition. But the time-lapse images do not reveal this information and are not perceived by viewers in the same way as continuous "real time" (i.e. the 30 frames per second of standard video) images. An imaging rate that rises above the flicker fusion level of about 20 frames per second, somewhat dependent on the brightness of scene illumination, produces a perception of continuous motion that is processed in the brain in a quite different way from a series of still pictures. Different information is extracted from a real-time series of images than from the still images, with some details being completely overlooked. It is possible to detect posture, rather than patterns of motion, in the series of still images, but this is of much less use for recognition.

The discrete images also worsen the effect of the noise. To some extent, it is possible to "see through" random or periodic noise when the images arrive at a rate above the flicker fusion level. When I was growing up, we lived far from major towns and the television picture that we could receive was heavily corrupted with "snow," random white specks that danced over the screen. But we could (and did) still watch Milton Berle and other shows, because the moving images of the real signal and the dancing snow of the noise moved at different rates of speed, and we could filter the one out from the other. When the image rate drops below the flicker fusion rate this ability is lost — the real image and the noise both move at the same speed and cannot be separated.

There are a few instances in which the actual velocity of motion can be measured from the still images. If the time between the successive images is known, as it usually is, then measurement of the distance traveled gives the velocity. Measuring the distance requires knowing the geometry of the scene and camera placement. The only common use of this technique is for measurement of vehicle speeds by roadside cameras, and in that case there is usually a set of lines painted on the pavement to assist in locating the vehicle in each image.

Finally, there is no sound recorded in surveillance video. This is partly because of the storage requirements, partly because of cost, and very much because it is not that easy to pick up clear sounds from large areas anyway. But if recognition is the goal, then the lack of sound is important. Most people can recognize persons they know by their voices alone, using a combination of the sound itself and the characteristic use of words and phrases. Eyewitnesses ("earwitnesses"?) at a crime scene often disagree about what was said, and there is no audio analog to the police sketch artist to characterize the voice of the perpetrator.

For all these reasons surveillance video is rarely able to provide unequivocal identification of perpetrators. Fortunately, in most cases it does not have to be used for this purpose. Showing a few images from the video, after modest contrast and brightness enhancement, on the evening news may be enough to spur acquaintances of the perpetrator who know about the crime to come forward with information. Showing the same pictures to the arrested suspect, who knows better than anyone else in the world what he or she looks like, may convince them that the images do constitute good identification and push them into confession or plea bargaining. In these cases, the video has done its job, and the images will never need to come before the court.

Enhancement

Several examples of image enhancement were shown in detail in Chapters 2 and 3. In principle, these are all applicable to digitized video images as well as to still images from a digital camera. The problem with surveillance video is the noise level present. All of the enhancement methods, even simple ones such as expanding the contrast and brightness to show detail in dark regions of a backlit image, also magnify the visibility of the noise. Some of the methods, such as unsharp masking, magnify the noise much more than the information content of the image. It is therefore of the highest importance to reduce the noise before any enhancement steps are taken.

With conventional video rather than time lapse surveillance recording, it is sometimes possible to perform temporal averaging to reduce random noise, by averaging together successive video frames. If the subject is not moving, or if it is possible to track the motion and shift the individual frames so that the images of the subject stay in registration, then adding the multiple images together produces an improved signal to noise (S/N) ratio. This ratio improves in proportion to the square root of the number of samples averaged, so adding together 4 frames would yield a factor of 2, and adding together 16 frames (slightly over a half second) would yield a factor of 4. This might then permit some of the more aggressive enhancement methods shown in Chapters 2 and 3 to be applied. With time lapse video this is rarely possible, however, as it would require the subject to remain stationary for tens of seconds.

In Chapter 2, spatial averaging was used when temporal averaging was impractical. This involved averaging together pixels within a neighborhood, to obtain better S/N values for uniform regions. In digitized video, it has been pointed out above that the resolution elements are larger than the

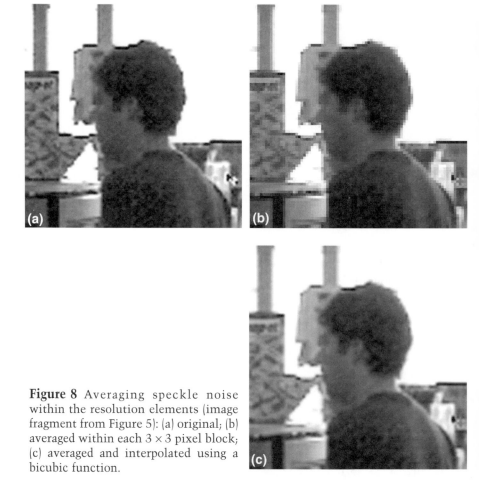

Figure 8 Averaging speckle noise within the resolution elements (image fragment from Figure 5): (a) original; (b) averaged within each 3 × 3 pixel block; (c) averaged and interpolated using a bicubic function.

captured pixels. It should be acceptable, therefore, to average together the pixels within each resolution element to obtain lower noise. As shown in Figure 8, this does in fact reduce the speckle in the image. Simply averaging the values within each resolution element creates a blocky or pixellated appearance that is visually distracting. Performing the same averaging but then smoothly interpolating values for the pixels in the display avoids this blocky appearance and certainly improves the visual appearance of the image. Indeed, this is the method sometimes used when surveillance video images are converted to hard copy printouts.

Note, however, that averaging between the doubled scan lines only corrects for noise introduced in the electronics and analog-to-digital conversion steps after playback of the video tape. This type of averaging does not deal with the major sources of random noise (the camera and video tape), nor with any form of periodic noise. The histograms of the images (Figure 9)

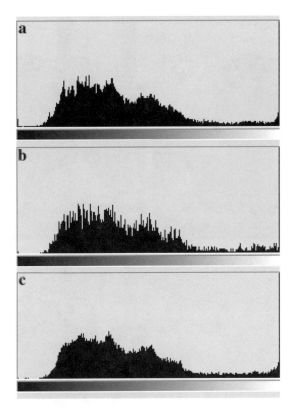

Figure 9 Histograms from the image fragments in Figure 8.

show only slight reduction in the widths of peaks, so the tonal resolution has not been improved by enough to permit much further enhancement. Figure 10 shows the application of an unsharp mask to the original and averaged images from Figure 8. While Figure 10c is visually more pleasing, it is not clear that any more useful detail has been recovered. It is a useful general rule that enhancement of an image only becomes worthwhile after any noise has been substantially reduced; enhancement techniques applied to noisy images serve only to increase the visibility of the noise and further obscure any useful details.

Image Restoration

One of the very aggressive enhancement techniques shown in Chapter 3 is deconvolution of the point spread function to restore a blurred image to sharper focus. This is done using the Fourier transform, and is quite sensitive to noise in the image. However, in cases of extreme importance when the

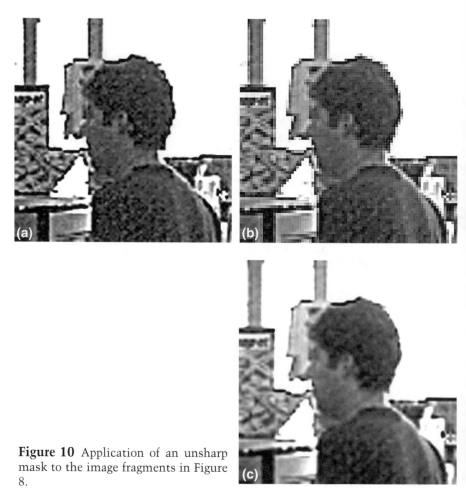

Figure 10 Application of an unsharp mask to the image fragments in Figure 8.

necessary information can be collected and image noise can be reduced, it may be justified for video images.

The first essential step is to remove as much noise as possible from the original image. This will not improve the spatial resolution, but it will improve the tonal resolution. If the noise is periodic, such as electronic interference, clock noise in the analog-to-digital converter, etc., removal can usually be accomplished quite well. If there is interlace jitter, the even and odd lines can be shifted to restore alignment of features and boundaries. Random noise can really only be improved by temporal averaging. For conventional video, even a few seconds of frames that show a stationary or uniformly moving target (so the frames can be shifted to bring the individual images into registration) may be adequate to average together enough frames to obtain a satisfactorily low noise image.

The most time consuming part of the restoration problem is character-izing the point spread function of the system. It is essential that the equipment be in the same condition and configuration as when the original images were acquired. In particular, lens settings and focus must not have been altered. If these conditions are met, then you should record a scene in which there is just one point source of light, with everything else dark. I have found that using a small "grain of wheat" flashlight bulb works well for this. The light should be quite bright, as it must be recorded at close to maximum intensity by the system.

The light source should be placed at the same location as the subject in the image. The assumption in deconvolution is that the same point spread function is present for the entire scene. In real cases this may not be true. Some parts of the scene may be at different distances from the lens, outside the depth of focus or subject to other distortions. The deconvolution method may actu-ally degrade the sharpness of some parts of the image for which the point spread function is incorrect, in the process of improving the area of interest. If there are several areas of interest (e.g., several people present in different parts of the scene) then it may be necessary to process each separately.

The image of the point source, after passing through the imaging system, will not be a single point. Unlike the case shown in Chapter 3, Figure 25, the point spread function from most surveillance systems will not be symmet-rical. Instead, as shown in the example of Figure 11c, it will probably have quite different characteristics in the vertical and horizontal directions, and also show a "comet tail" that smears out the light in the scan direction. These effects result from many different parts of the optical system, including lens flare or out-of-focus optics, light scattering and electron tunneling in the chip, light continuing to fall on the array during readout, inadequate band-width in the electronics, and of course a multitude of possible effects in the video recorder.

The deconvolution process divides the Fourier transform of the scene (after noise reduction) by the Fourier transform of the measured point spread function. This is complex division, since each value in the transforms has both a real and imaginary part. The resulting data is then inverse Fourier transformed to reconstruct the spatial domain image, from which the blur has been removed. As shown in Figure 11, this makes the information in the picture much clearer (in this case it enables the license plate to be read without difficulty).

Notice that the image in Figure 11d has increased noise content. This is the typical effect of deconvolution, because division of one value by another always increases the effect of any random variation (noise) in the values. Classic deconvolution, as shown in the example of Figure 11, is very sensitive to the effects of noise in the original image. The reduced noise image in

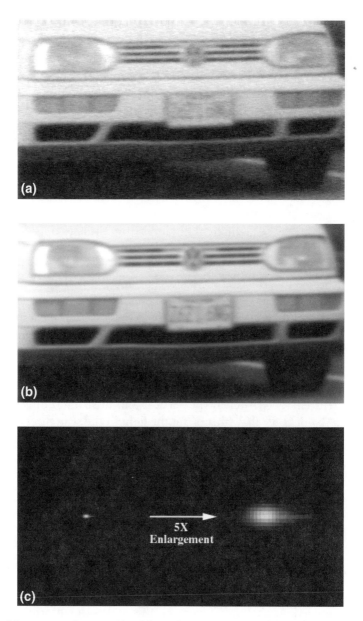

5X
Enlargement

Figure 11 Deconvolution of a blurred, noisy image: (a) one video frame; (b) average of eight frames, showing reduced noise; (c) point spread function of the imaging system, as described in the text; (d) deconvolution of image (b) using image (c); (e) deconvolution of image (a) using image (c).

Figure 11b that was used for deconvolution was obtained by adding together eight video frames, each similar to the one in Figure 11a. The image of the point spread function must be acquired under identical conditions as the

FORENSIC USES of DIGITAL IMAGING

Color Plates

Plate 1 Image of an evidence scene with a color scale included. (See Figure 1.11.)

Plate 2 Enlarged view of a printed halftone color image. (See Figure 1.18.)

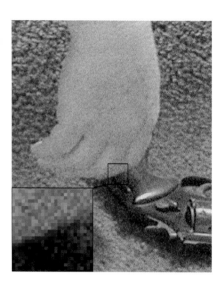

Plate 3 Fragment of a color image enlarged to show pixel noise: original (compare to Plates 4, 15, and 16). (See Figure 2.8a.)

Plate 4 The image in Plate 3 with noise reduced by processing individual RGB planes as described in the text. (See Figure 2.8b.)

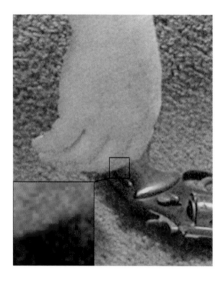

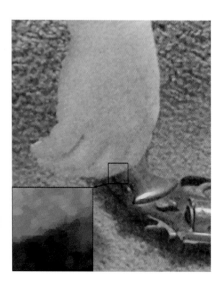

Plate 5 The image in Plate 3 with noise reduced with color median as described in the text. (See Figure 2.8c.)

Plate 6 Removal of pattern noise due to detector variations by subtracting a blank image: original image (enlarged), showing colored pixels resulting from 8-second exposure time (compare to Plate 7). (See Figure 2.10a.)

Plate 7 The image in Plate 6 after subtracting a blank image collected for the same time. (See Figure 2.10b.)

Plate 8 Image of a steer standing in a field (compare to Plate 9). (See Figure 2.19a.)

Plate 9 The image in Plate 8 showing the effect of rotating the hues by about 25 degrees. (See Figure 2.19b.)

Plate 10 Image of cars in a parking lot, as photographed (compare to Plate 11). (See Figure 2.20a.)

Plate 11 The image in Plate 10 after adjustment of R, G, and B intensities to make the paving neutral grey. (See Figure 2.20b.)

Plate 12 A portion of a color chart: original image (compare to Plate 13). (See Figure 2.21a.)

Plate 13 The image in Plate 12 after tristimulus correction using the values in Chapter 2, Table 3c. (See Figure 2.21b.)

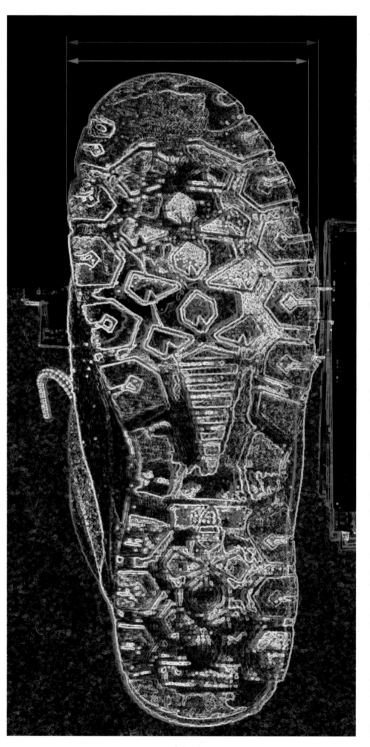

Plate 14 Superposition of the boot sole from Chapter 2, Figure 25 onto the footprint image of Chapter 2, Figure 23, using color channels. (See Figure 2.26b.)

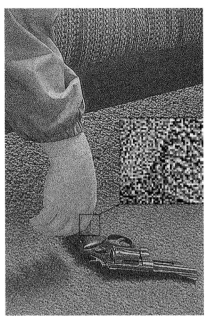

Plate 15 Sharpening a noisy color image. The image from Plate 3 after sharpening the individual red, green, and blue channels. (See Figure 3.6a.)

Plate 16 The image from Plate 3 after applying an unsharp mask (standard deviation 5 pixels) to the intensity only. (See Figure 3.6b.)

Plate 17 Color image: original, with high contrast and little visible detail in shadow areas (compare to Plate 18). (See Figure 3.12a.)

Plate 18 The image from Plate 17 after adaptive equalization, showing enhanced detail and improved visibility in shadow areas. (See Figure 3.12b.)

Plate 19 Color filtering: original image of a check (compare to Plate 21). (See Figure 3.16a.)

Plate 20 Selection of a yellow filter complementary to the blue lines in Plate 19. (See Figure 3.16b.)

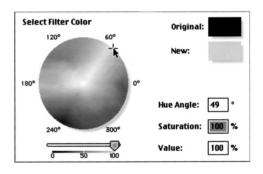

Plate 21 The image from Plate 19 with the application of the yellow filter. (See Figure 3.16c.)

Plate 22 Separation in HSI space: original color image. (See Figure 3.17a.)

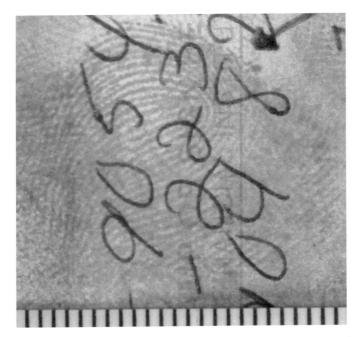

Plate 23 Expansion of HSI range: original image (compare to Plate 24). (See Figure 3.18a.)

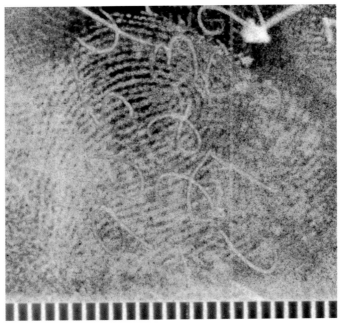

Plate 24 The image from Plate 23 with histograms for the H, S, and I channels set to expand their range. (See Figure 3.18c.)

Plate 25 An original color image (used for filtering and compression). (See Figures 3.15a and 3.34a.)

Plate 26 Reconstruction of the image from Plate 25 after 38:1 wavelet compression. (See Figure 3.34b.)

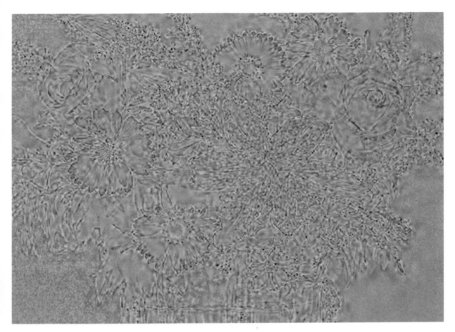

Plate 27 The difference between images in Plates 25 and 26. (See Figure 3.34c.)

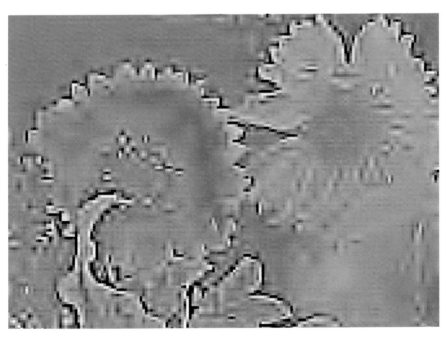

Plate 28 Application of a Laplacian sharpening operator to the image in Plate 25 (enlarged to show detail). (See Figure 3.34d.)

Plate 29 Comparison of two attempts to disguise a human face: original image (compare to Plates 30 and 31). (See Figure 4.15a.)

Plate 30 The image from Plate 29 with replacement of eyebrows, ears, hair, nose, and mouth, and altering eye color. These do little to disturb recognition. (See Figure 4.15b.)

Plate 31 The image from Plate 29 with small alterations in height of the forehead, distance between the eyes, depth of the chin, width of the jawbone, width of the nose, and protrusion of the ears. These make the face appear very different. (See Figure 4.15c.)

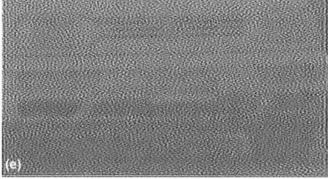

Figure 11 (continued)

original blurred scene, which in this example means that the same temporal averaging was used for Figure 11c. If the deconvolution was applied to the noisy image in Figure 11a the results would be completely dominated by noise, as shown in Figure 11e.

In those cases in which the noise content of the image can be estimated, if not removed, it is still possible to perform a partial restoration. Instead of dividing by the FFT of the point spread function, a constant that is a function of the noise content of the image is added to the divisor. This is Wiener filtering, covered in detail in most image processing textbooks.

In deconvolution, it is also important as a practical matter of implementation that the division be limited to those areas where the magnitude of the transform of the point spread function is not too small, to prevent numerical overflow, and to perform all of the arithmetic calculations with high numerical precision, not the 8 bit integer values typically used to hold the original image, but double precision floating point values to preserve the contributions of all of the terms in the Fourier transform.

Recognition and Identification

Recognition and/or identification of faces and features from images is difficult under the best of circumstances (which surveillance video does not provide). It may be helpful to distinguish between recognition and identification. In the following discussion recognition will be used to denote the process by which we realize that an image or face is one that we have seen before, while identification will be used to denote the process of connecting a specific name to a face, or determining that one image matches another, or matches a specific person. These are different processes. Many of us have, at one time or another, recognized a face as familiar, but been unable to recall who the person is (and, indeed, many of those "recognition" events turn out to be incorrect).

The reliability of eyewitness identification of persons is notoriously poor, even when the encounters are live. With video that lacks color, has limited spatial and tonal resolution, and lacks sound or continuous motion, the problem is obviously worse. There is a great difference between confirming that an image matches (or could match) someone whom we know well and expect to see in the picture — whether this is a familiar weatherman on the 11 o'clock news or a family member in a home video — and being able to match the image with someone who is not familiar.

The interesting thing about recognition, as understood by perceptual psychologists and others working with human and computer recognition, is that there are two entirely different mechanisms at play in different circumstances.

Unconscious recognition is often modeled by something like the Perceptron introduced as a key tool in computer-based artificial intelligence half a century ago and now, in somewhat modified form, used in neural net logic. A large number of inputs are tested against stored criteria. For instance, if we consider the problem of recognizing someone familiar (this is usually called the "grandmother cell" scenario), there are lots of clues that we would associate with them. Hair and eye color, facial features such as wrinkles or dimples, the relative spacing of eyes, nose, and mouth, etc. It is believed that the brain has evolved a lot of specialized tools for recognizing faces and facial features, many of them operational from the moment of birth: tiny children track faces and eyes.

Clearly, most faces come with the same number of eyes, ears, noses, etc., so it must be secondary factors that are involved in recognition. As discussed below, the shape and, in some cases, the size of these features can be diagnostic, but as we have seen above, video images do not generally offer enough spatial resolution to capture these factors. Placement of features relative to each other may be more important. True, everyone has their nose above the mouth and between the eyes, but the spacing of these features may vary. There is enough variation in these spacings, and the presence or absence of secondary features such as wrinkles, dimples, moles, etc., to assist in making identification. In some cases, video images may have enough spatial and tonal resolution to support the detection of these characteristics.

On this basis, the visual system is always examining images of faces and extracting the key features (distinguishing marks and feature placement), and passing them to a "grandmother cell" for analysis. The observed factors are compared to the remembered factors for grandmother. Those that agree generate positive impulses, and those which disagree generate negative ones. Factors that are not observed (e.g., you may see someone from the side and not be able to check eye color) are ignored. It takes only a few positive impulses (hair color, the correct eye spacing, etc.), in the absence of any negative ones (the presence of a red mustache would presumably eliminate the face as a candidate for grandmother) to trigger a "recognition" signal. If there are a few negative matches it takes many more positive ones to overwhelm them, which is the basis for most attempts at disguise.

The method is very fast, works always and automatically, and is responsible for our ability to recognize grandmother in unfamiliar or unexpected settings, and from only partial information. It also results in a high level of mistaken recognitions and incorrect identifications, when we think we see someone familiar but as more information is received and more matches can be tested realize that it is not the person we first thought.

The number and variety of features can be judged by the ways that police sketches are generated by software. Only a small number of shapes for noses, eyes, ears, etc. need to be stored. These can be stretched in size and moved in position to generate sketches that match people's recollection of faces they have seen. One of the drawbacks of face recognition is that we generally do not know which features and spacings are most important for people of racial or ethnic backgrounds that are unfamiliar to us. Apparently, the key factors are different for different population groups, which is the reason members of one group "all look alike" to someone from another group.

Conscious or deterministic identification uses the same kind of logic, but in a different way. In this case we are intentionally scrutinizing a face to see if it is the person we expect, or of whom we have a picture (either a physical picture or one stored in the mind). In that case, negative matches (some feature that is expected but not observed, or some feature that is observed that is not present in the picture) are given high weight: they essentially rule out identification. But positive matches are not much more important than ambiguous ones. In other words, any generic image of a face would match any person, unless there was some clear conflict between the image and the actual person's face.

This is unfortunate for the use of surveillance video images, because they are usually quite ambiguous. There are few clues that the images distinctly reveal, perhaps only the length of hair (on the head or face, sideburns, whiskers, etc.) and the presence of major distinguishing marks. Thus the video image could be almost anyone we want or expect it to be. The person has to be clearly different from the low-resolution ambiguous image before we fail to accept the possibility that they are the same.

Illustrations such as that shown in Figure 12 are sometimes cited as evidence that human recognition requires very little information and that very low resolution images can be useful. Actually, it is evidence for something quite different, but also important. Viewers who "recognize" Abraham Lincoln in the blocky image do so based on very little information, it is true, but is the image really that of Lincoln? The image pixels are ambiguous and could be any of many people. In fact, we are quite likely to identify an image as someone famous, or familiar, or simply in front of us (e.g., every day as the defendant in the courtroom) provided there are no obvious differences between the person and the picture. The more ambiguous the picture, the less likely that those differences will be found.

Figure 12 A block image representing Abraham Lincoln, sometimes used as a pattern for quiltmaking.

Identification of Faces

The human brain is programmed by evolution to detect images of faces. Even tiny babies instinctively locate and track faces in their field of view. The components that "make up" a face (primarily eyes and mouth, as seen in the simple "happy face" cartoon in Figure 13) occur often enough in nature that we tend to see faces even in the accidental placement of natural objects and shadows. The "Mars Face" (Figure 13) is merely a recent example of this phenomenon.

However, while the existence of a "face" image requires only a few principal marks, the recognition and identification of faces seems to depend on the correct placement of features within the face. The major dimensions that are used are shown in Figure 14. The vertical and horizontal spacings are combined as ratios, so that the absolute size of the face or image is not critical.

The importance of these dimensions as compared to the fine details of the face image itself can be demonstrated easily. In Figure 15, a face image has been modified by replacement of the hair, ears, eyebrows, nose, and mouth and recoloring of the eyes. It is still recognizably the same person, and indeed the individual had a difficult time in identifying all of her own features that had been replaced in the picture. On the other hand, keeping all of the original facial details the same, but slightly altering the dimensions (height of forehead, protrusion of ears, separation of eyes, size of chin and width of jaw, width of nose) produces a face that is very difficult to recognize or identify as the same person. The art of makeup and disguise utilizes such changes in the shape and dimensions of the face (e.g., produced by internal pads and applied surface layers of rubber) very effectively.

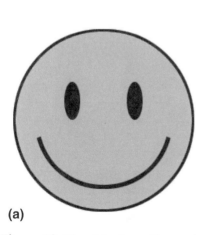 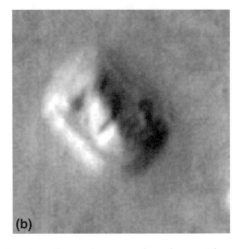

(a) (b)

Figure 13 The ubiquitous "happy face" cartoon has only a mouth and eyes. The "Mars Face," an accidental grouping of shadows with the same mouth and eyes, is also perceived as a face.

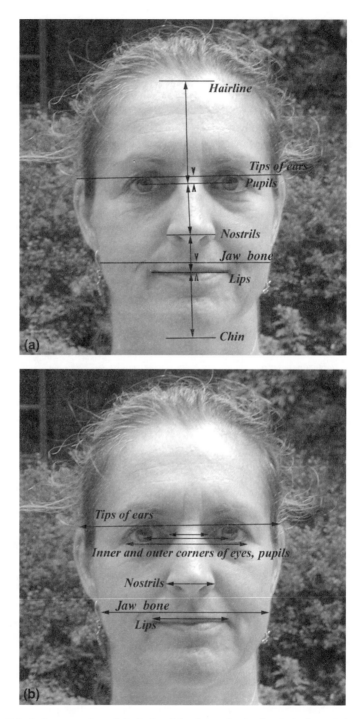

Figure 14 A human face labeled with the principal vertical and horizontal dimensions used in facial identification.

On a finer scale, the presence, shape, and placement of distinguishing marks such as moles, wrinkles, smile lines, crows' feet, and other folds and blemishes on the skin can have a role in identification. But in most cases it is only the presence of an unexpected blemish or mark that is noticed, and considered as a barrier to matching an image to a face. In any case, there are many fewer distinguishing marks of identification in a facial image than in, for example, a fingerprint, particularly as people tend to overlook many of them. Law enforcement officers are trained to look for "distinguishing marks" such as scars and tattoos.

The human ear is rich in possible identifying details. The shape and dimensions of the ridges of cartilage within the ear and of the ear lobe vary among individuals, although with fewer unique points of identification than a fingerprint. However, ears are rarely seen at an advantageous viewing angle in surveillance photographs (see Figure 6 for an example), and may be obscured by hair or a hat. Of even greater concern is the inadequate resolution of most surveillance images to reveal the important structure of the ear. As shown in Figure 16, a resolution of about 1 mm is barely adequate to show the details of the ear, and a resolution of 2 mm is not adequate. Surveillance images rarely show heads large enough to produce resolution of 1 cm, so that the details of ear structure are not often useful for identification.

Identification by DNA

Another important technique for identification in many situations is DNA evidence. Most of the process for handling this evidence has nothing to do with imaging, and so will be reviewed here only briefly. DNA can be extracted from almost any human tissue, so that DNA sources found at a crime scene may include blood, semen, tissue, hair follicle cells, or saliva. DNA extracted from items of evidence is compared to DNA extracted from reference samples taken from known individuals. DNA evidence is also important in establishing paternity.

Extracted DNA is treated with an enzyme that cuts the double stranded molecule whenever a specific base pair sequence occurs. The length of fragments between the same markers differs from individual to individual because of the variable number of random repeats of introns — segments of the DNA molecule that are believed to have no useful genetic function. Following this digestion, the DNA fragments are separated by electrophoresis in agarose gels. The molecular fragments, which are negatively charged, migrate toward a positive electrode. Smaller fragments move more rapidly through the pores of the gel matrix than larger ones, producing a separation of the DNA fragments according to size. Following electrophoresis, the

Figure 15 Two attempts to disguise a human face: (a) original image; (b) replacement of eyebrows, ears, hair, nose, and mouth, and altering eye color, does little to disturb recognition; (c) small alterations in the height of forehead, distance between eyes, depth of chin, width of jawbone, width of nose, and protrusion of ears, make the face appear very different. (Shown as Plates 29 to 31.)

separated DNA fragments are denatured (the DNA strands are separated) while still in the agarose gel by soaking in a basic solution. This leaves single strand DNA fragments which are transferred to the surface of a nylon membrane by blotting, a procedure known as a "Southern Blot" which preserves the spatial arrangement of the DNA fragments.

The blot is then hybridized with a probe chosen to detect one genetic locus or sequence on a single human chromosome. The probe is tagged with a radioactive label, and pairs with the DNA molecule where it is complementary to its base sequence. This labels the band in the column that has that marker, and whose location identifies the size of the fragment from that individual. After hybridization, the unbound probe is washed away, so that the only radioactivity remaining on the membrane is associated with the DNA of the targeted locus. The locations of radioactive probe hybridization are detected by placing the membrane next to a sheet of X-ray film or coating it with a film emulsion to record the radioactive decay. The developed X-ray film is called an autoradiograph. After an autoradiograph has been developed for the first single locus probe, the radioactivity is washed away with a high

Figure 15 (continued)

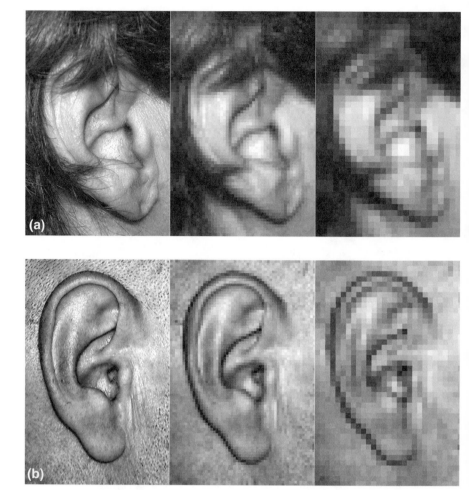

Figure 16 Two different human ears. In each set, the high resolution image at the left has been reduced to a resolution of 1 mm (center panel) and 2 mm (right panel). At 1 mm resolution, the shape and dimension of the distinguishing ridges and lobe are barely discernible. At 2 mm resolution they cannot be used for effective identification.

temperature solution, leaving the DNA in place. The blot can be hybridized with several radioactive probes by repeating the steps above. The complete autoradiograph is known as a DNA profile, and sometimes referred to as a DNA fingerprint.

The developed X-ray film that records the location of the various hybridized probes typically contains the results for several different samples arranged in side-by-side columns on the gel, allowing direct comparison of the DNA fragments from the crime scene and various suspects. The appearance of dense bands at the same locations (distance from the starting point) constitute matches. Since each polymorphism is present in some fraction of the population, which is known at least approximately, the probability that a particular sample of DNA that matches all of the bands from a suspect could have come from someone else declines rapidly as the number of probes is increased. As an example, if six probes were used and each targeted (and matched) polymorphism was present in 5% of the population, the probability of a false match is $(0.05)^6$ or one in 64 million. Because the frequencies of occurrence are not well known in all population and racial groups, and may not be entirely independent, these probabilities are somewhat controversial, but in any case are very low. On the other hand, if even a single band from the evidence is not present in the suspect's DNA, or vice versa, then that person is not matched and can be excluded from the suspect list.

Normally, the DNA from crime scene evidence and that from various suspects are processed side by side in the same electrophoresis gel, so that direct comparison of the bands in the different columns is possible, as shown in Figure 17. Even in a single gel, there are often small variations that cause migration rates in different columns to be slightly different, but this does not normally cause any difficulties. Making comparisons between different gels requires adding marker proteins of known weight to each sample. These migrate in the gel along with the DNA to produce a set of fiducial marks that can be used to normalize the distances of migration and permit comparison. Such markers, which are usually placed in one or more separate columns on the gel, are an essential component of efforts to establish a database of DNA markers that could be used just as fingerprint files (and to a lesser extent mug shots) are used now, to identify a suspect from evidence found at the scene of the crime.

That is where digital image analysis enters the story. Autoradiographs can be digitized with a flat-bed scanner just like any other film transparency. But storing the images themselves requires large amounts of storage, and does not lend itself to matching. Instead, a set of numbers recording the position of the bands for the various probes is needed. Computer-based image analysis is commonly used to convert the image to a set of measurements of band position, and perhaps band density. The position of each band can be measured by ratioing to the distance of the marker proteins, although this must be done with some care because positions along the columns are not necessarily linear. Much more troublesome is the question of how to determine whether a band is present. The fit of the probe to the DNA does not have to be exact; a poorer match between the sequence of base pairs forms fewer hydrogen bonds and consequently the density of the band is lower in the resulting autoradiograph. This also varies with the temperature of the hybridization reaction, and the salinity of the chemical environment. Variations in film exposure and development also cause variations in band

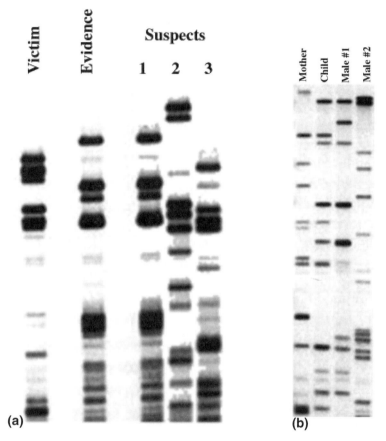

Figure 17 Two example DNA autoradiographs: (a) comparison of DNA fragments from crime scene evidence with the victim (in order to exclude any bands
which match due to possible contamination with the victim's DNA) and several
suspects; (b) comparison of DNA fragments from a child with those from the
mother and two males (the child's DNA comes from both parents and so each
band must match one parent or the other).

density. It is not easy to decide whether a low density band represents a match
that should be included in such a database or not. Of course, once a suspect
is identified and located, a direct comparison can be made between the crime
scene evidence and a DNA sample to confirm the match.

Fingerprint identification has been used for most of the twentieth century
as the most reliable means of unequivocally identifying individuals. Fingerprints are the patterns of ridges and valleys on the fingers, palms, and soles.
In addition to the obvious large scale structure, which is described as a loop,
whorl, or arch (Figure 18), these patterns contain specific features called
minutia, which are bifurcations or branchings in ridges, and ridge endings
(Figure 19), which are unique and allow specific identification of individuals.

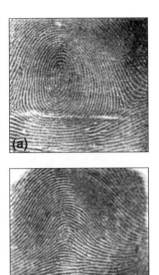

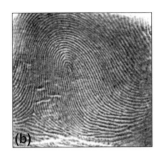

Figure 18 A few of the standard fingerprint images compiled by the National Institute for Standards and Technology (NIST), showing typical structures: (a) whorl; (b) loop; (c) arch.

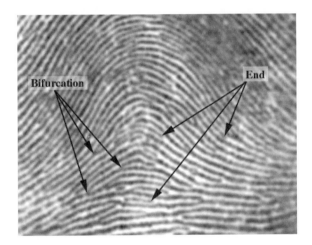

Figure 19 Minutia in a fingerprint consist of bifurcations and line ends.

The minutia are recorded with three coordinates: the x, y location of the branch or end, and the angle a which represents the average orientation angle of the lines near the minutia point. Indexing of this information from a fingerprint image is done automatically using sophisticated software for high quality images such as ten print cards, in which the image contrast is high, there is little blurring or smudging, the entire finger is recorded "nail-to-nail" by rolling the finger on the card, and the angle of the finger with respect to the card is controlled. There are also optical fingerprint scanners used to obtain fingerprints directly from a hand, used in some security applications. For latent prints obtained at a crime scene the image is usually indexed manually since the images are often of poor quality and may be incomplete.

Matching of the data for minutia from an unknown "search print" against a database of "file prints" is performed by software that attempts to superimpose the two sets of points, allowing for shift and rotation, and looks for clusters of matches. Each match is given a score based on the number of minutia matched; the highest scores correspond to the most likely matches. From each search, the few (typically 10 to 100) highest scoring matches are then reviewed by a human specialist who examines the actual stored images. These are compressed using a wavelet compression routine that takes advantage of the fact that the ridges in human fingerprints are quite uniform in spacing, and higher and lower frequencies in the image can be eliminated without affecting the important information. Final human judgment is considered essential because prints can be distorted, blurred, or smeared, and are not necessarily complete. Matching of ten print cards against the 27 million file prints in the IAFIS (integrated automated fingerprint identification system) operated by the Federal Bureau of Investigation is considered quite reliable, and can process approximately 8000 searches daily. Matching of a single latent print, which may come from any of ten fingers, is much more difficult, operates with a smaller database, and requires much more human involvement by specialists.

Misdirection and Mistakes

The clever criminal would prepare for a robbery or other crime in which surveillance video might be present by arranging to provide it with false distinguishing marks. A realistic wig to change hair style or produce a bald spot, cheekpads to change the shape of the face, the addition of sideburns and facial hair, a few prominent scars, perhaps contact lenses to change eye color, and dramatic clothing and perhaps a prominently bandaged arm, and a watercolor "tattoo," all clues that could be discarded immediately afterwards, would provide a host of false matching factors that would overwhelm attempts at human recognition and identification even if the images were of much higher quality than surveillance video. Fortunately, most criminals (at least those that are caught) are not very bright and don't know much about the logic of recognition and identification, or about the resolution of surveillance imagery.

The same phenomenon of ambiguous images is well known in mug shot collections, which have much higher resolution than video images (although often not as much resolution as a typical modern digital camera would

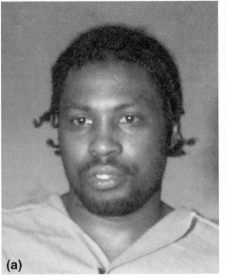
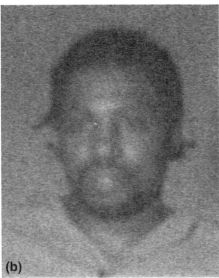

(a)   (b)

Figure 20 Comparison of the same image with typical photographic resolution (either film or digital) with surveillance video, showing reduction in resolution and superposition of noise.

provide, along with the convenience of internet availability). Figure 20 shows a comparison between the high resolution of a photograph as obtained with a typical film or digital camera, as compared to the same image degraded to the typical resolution and noise content of a surveillance video image (although the face is still a much larger fraction of the image than it would be in a typical surveillance situation). When viewed side by side, it is possible to find similarities between the two images. The noisy, low resolution image does not, however, contain enough details to be used for identification of the subject. Even for the photographic quality image, many pictures may be accepted as possibly matching a face.

In a famous example of this type of confusion, a prisoner named Will West was being sent to Leavenworth prison in 1903. But the clerk at the prison insisted that he was already there, serving a life sentence for murder; and indeed there was a prisoner of that name already incarcerated there. Comparison of mug shots (reproduced in Figure 21) showed that they were sufficiently similar in appearance that the photographs (as well as measurements of height, cranial shape, etc., an identification system now long discarded) could not distinguish them, although they were unrelated. Based on a few such instances, in which serious miscarriages of justice could have resulted from mug shot identifications, fingerprint identification came into general use as the only reliable means of identification (at least before widespread use of DNA typing is undertaken).

Figure 21 The two Will Wests.

Video imagery, especially surveillance video with its poor spatial and tonal resolution, and less than optimum lighting and viewing angles of faces, is worse for identification than mug shots.

Note that the feature matching method described above is not the same way that computer systems recognize or identify faces. Newly developed systems for use in bank cash machines, employee identification, and other similar "secure" installations are expected to replace PIN numbers ("personal identification numbers") for identification. They select a few features on the face and measure them carefully. Some use the dimensional ratios shown above, but the most successful example to date of an automated security system based on video image analysis uses the pattern of random tears that form after birth in the iris, around the pupil of the eye. Measuring the pattern of light and dark texture by image analysis, as shown in Figure 22, provides a signature as unique as a fingerprint (taking into account the influence of eye rotation, reflections, the presence of glasses, etc., the claimed error rate is less than one in a million) and not susceptible to change, as hair style or makeup can be used to alter other characteristics of appearance. Of course, this is not something that a human observer would use for identification nor something accessible from normal video surveillance imagery.

The level of protection required from these automated identification machines is quite high. It must at least match that afforded by three guesses at entering a four-digit PIN number, or one error in 1333 cases for random guesses. This is the typical level of security in current generation cash machines, and is apparently the rate of false matches that banks are willing to tolerate before handing out a few hundred dollars. In court proceedings

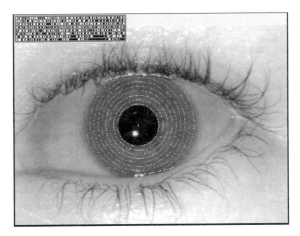

Figure 22 Coding the pattern from the iris of a human eye (courtesy IriScan, Inc.).

in a capital murder case we require a much lower rate of certainty: "beyond a reasonable doubt" is often described as meaning 90% certainty, which translates to one error in 10 cases.

The Expert Witness

5

Working with Counsel, Judges, and Juries

Expert witnesses may be called in many fields, only some of which involve science. In the case of image analysis, there is a scientific basis that underlies the testimony. This may possibly involve such diverse fields as mathematics, computer science, optics, electronics, the psychology and physiology of human vision, microscopy, and of course whatever fields are involved in the analysis of the evidence shown in the images. The expert is supposed to be able to assist the decision makers (the jury) in understanding and interpreting the evidence. But the expert is hired by one of the advocates (counsel) and not by the court itself. This is not the practice in many European countries, where expert witnesses are brought into the case by the judge, but it is the situation in common law countries such as England, Canada, and the United States.

It is not unusual for both sides in a case to retain their own expert witnesses, who will be asked to testify by answering questions asked by counsel and will often contradict each other, or at least appear to do so, because of the particular questions that are asked and the order in which the evidence is presented and interpreted. Also, the advocacy system practiced in our courts is adversarial, so each counsel will not only try to elicit the most favorable testimony from his or her experts, but will try to minimize and impugn that from opposing witnesses.

This makes many expert witnesses, who are usually trained scientists, very uncomfortable. They expect in their disciplines that the push and pull of ideas and hypotheses will be settled by the collection of facts, and that truth will emerge. I've been reminded more than once that trials aren't about truth, but about justice, which is apparently something else altogether.

Scientists further expect that their hypotheses and opinions, supported by logic and scientific knowledge, will be respected and that they will not personally be attacked. But counsel will often try to rattle an opposing expert by selective misquoting, taking statements out of context, trying to trap the witness into contradiction or mistatement, and often by personal attacks that have nothing whatever to do with the testimony.

Of course, no one enjoys being confonted by a barrage of questions that are often irrelevant, repetitive, sometimes personal in nature, rude, and moreover come from someone who quite evidently does not know as much about the science as the expert does. It is tempting to show a flash of anger, put the rude person in their place, and claim the moral and scientific high ground by hautily pointing out that the expert is, after all, the expert. To do this is almost invariably a bad idea, and in fact just what the opposing counsel is hoping will happen. The expert witness who can remain calm and cool in the face of this type of questioning and who can even turn it to an advantage is the one who succeeds.

The only functional response that I know, which usually works, is to remember that your role is to help the jury understand the evidence. So don't make your reply to the questioning attorney. Talk to the jury. Treat them as people who may not know all of the underlying science, but who are certainly capable of understanding and appreciating it if you can explain it clearly. Don't be afraid to reiterate what you said before (correcting the misquoted words that counsel is probably trying to put into your mouth). If necessary, something like "Let me take a minute to try to answer your question using a different explanation (analogies are especially good for this), since it is evident to me from your asking it again that I must not have done an adequate job of making myself clear the last three times." That usually tips the jury off as to what is going on, and also lets you get back to the education aspect of your testimony. It also signals the attorney that you probably aren't going to get rattled or confused.

Cross examination is a test of how quickly you can think on your feet. The best scenario is to turn the question around and use it to make your own points again, using the other side's own question (either with another image, or a hypothetical situation). Expert witnesses who are good at this kind of fast thinking, clear delivery and building rapport with the jury can be quite effective. Using such a witness early in the presentation of a case can equip the jury with a filter through which they view all subsequent evidence. On the other hand, some attorneys like to save the expert witness to the end, like the cleanup hitter in baseball, hoping that the last impression left with the jury will be the important one.

An example: in a trial in which the guilt of the defendant hinged critically on a surveillance videotape of very poor quality that showed the perpetrator's face with very little spatial and tonal resolution, I was asked in cross examination to examine enlarged and enhanced copies of several images. The cross-examining prosecuting counsel was intent on convincing the jury that from the series of pictures there was more information than in any one, and while one picture might not be a good match for the defendant the impression from the series of pictures should allow them to convict him. When the series of images was on the video monitors in front of the jury, I noticed something that made my point very well and undermined the cross examination.

I pointed out that on the three images shown from three consecutive video frames, because of noise on the taped image there were three entirely different shapes for the perpetrator's ear. In one frame it was smoothly rounded, in a second it had a definite lobe, and in the third it appeared unusually small. I pointed out these differences and asked whether it was reasonable to choose the shape ear that most closely matched the defendant and use that as evidence against him, or to choose the shape ear most dissimilar and use that as evidence for him. My answer was that neither was permitted - the only fair conclusion was that considering the quality of the tape, there was no information about ear shape (or any other detail of facial recognition) that could be dependably used. The images were ambiguous and therefore the entire tape was not to be treated as useful evidence. Not only had I refuted the specific point the cross-examining counsel wanted to make, but I had reiterated the point that I had tried to make in direct examination even more strongly, and hopefully left a lasting impression of that point with the jurors.

During cross examination, attorneys like to pose complicated multipart questions with hypothetical conditions, and then insist on a simple yes or no answer. Don't fall into that trap. You have the absolute responsibility to object to the wording of the question or to rephrase the question yourself to clarify any vagueness and correct any mis-statements, and then to follow your answer with a full explanation. If the attorney tries to cut you off by starting another question, just point out that you haven't fully answered the last one yet. Juries usually don't like overly agressive questioning, and will react against an attorney who seems to be trying to trick the witness (especially if it doesn't work). A lengthy and pointless rehashing of the same ground will probably bore the jury so that they will forget the main point being made (or challenged); of course, this in itself can be the aim of the line of questioning.

When the cross-examination consists of seemingly endless repetitions of the same, slightly reworded question, or grinds ever onward toward a point that you as the expert know is either trivial or (worse) a potential weakness in your case, it is tempting to resist. If you become caught up in a battle with the inquisitor you make two mistakes (which of course is the goal of that line of questioning). First, your attention wanders from the jury, whom you are supposed to be educating, to the opposing counsel. Second, you are apt to be perceived as a recalcitrant, biased witness who favors one side and hence is not the fair and unbiased expert that the jury can trust. It is vital that you exhibit no prejudice. So instead, when the line of questioning takes this tack you must work even harder to (a) keep your manner calm and your attention focused on the jury, and (b) find ways in your answers to work in the points that you wish to make and relate the explanations back to those points, even (or especially) when it involves repetition of things you've said before. You can tell if you are successful if the cross-examiner starts skipping ahead through their yellow legal pad of potential questions to try to find some other avenue to discredit you, because that one isn't working. If things get sticky, you can hope that the attorney who hired you will start objecting to the questions to give you time to settle yourself down.

You may occasionally encounter a witness for the other side who has been brought in as an expert but in fact isn't. Their only knowledge will be textbook or made-up examples, and they generally will not really understand why the examples they are proposing do not apply to the case in hand. The person may be sincere, but inexperienced or poorly trained. This is an all too common example of what is usually called "junk science." It may be your role to assist the counsel who hired you to detect the errors, omissions or other problems in the testimony of this so-called expert and to craft questions for cross examination that will reveal them. This is again an uncomfortable role for many scientists, who prefer to let ideas speak for themselves and rely on the accumulation of facts to squelch nonsense. Think of it again in the role of aiding the jury. They cannot (except in rare situations) question the witnesses themselves, and might not know what to ask if they could, so you as their teacher must help them by asking (through the counsel who hired you) the questions that will expose their lack of understanding or the errors in their testimony.

The other role of the expert witness is beforehand, to educate the counsel and help them to prepare questions for your testimony that will elicit the information most helpful to the case. Remember that you are not being asked to present all of the scientific evidence on both sides of the case. If there are weaknesses in your side's arguments, you should point these out to your counsel so that they do not ask questions that will expose those weaknesses. If the other side asks about it, then you are obliged to answer. But you do not have a responsibility to volunteer the information to the other side or to the jury. A trial is emphatically an adversarial procedure in which not all of the evidence will be presented. Each side seeks to present that which is favorable to its case, and to bypass and if possible exclude evidence that is unfavorable.

Counsel can try to exclude evidence on many grounds. These can include its being improperly collected in violation of a defendant's rights. Exclusion can also be based on the evidence being prejudicial, confusing, or unnecessary. Prejudicial evidence might include, for example, photographs of a victim showing horrible injuries that are unsettling. It is for this reason that autopsy pictures are almost never shown in a courtroom — they are simply repugnant to juries (and everyone else except for trained medical personnel) and might make them more likely to convict someone. Confusing evidence may, in the opinion of a judge, be anything that is too complex to help the jury reach a decision. Unnecessary evidence may include things that have already been established, or that seem not to have a direct bearing on the decision at hand, and would simply be a waste of time. It is up to counsel to convince the judge that evidence is not confusing and is necessary.

The dependence on the judge's decision to exclude offered evidence from presentation to the jury places a further requirement on the expert witness: it may be necessary to educate the judge about the meaning of the evidence and your opinion on it. Rarely will it be possible to do this directly. Judges usually restrict their input to what the opposing counsel present to them, in the form of arguments or questions to the expert witness. A few judges will ask questions directly of the expert witness in order to help them resolve specific points of concern. None in my experience will ask the witness to simply explain the relevance of the evidence to them, so they can decide whether it should go before the jury. The only way to work around this barrier is to make sure beforehand that counsel understands the issues and has the proper questions ready to ask you so that the important answers are expressed.

Scientists think of evidence as all facts and observations that may conceivably bear upon the truth of an hypothesis. In a court of law, evidence is whatever the judge allows to be presented to the jury. Much of the debate, particularly when scientific issues and (in the specific circumstances of interest in this book) images are involved, may take place in arguments with the jury not present as opposing counsel seek to convince the judge to include or exclude certain evidence from use. It is unfortunate that our system does not normally allow the judge to ask the expert for unbiased opinions on these issues. There are some exceptions to this in federal courts, and particularly in civil cases involving patents, but it is very rare in criminal trials. The only role of the expert witness in this instance is to provide assistance to counsel, who must turn the technical arguments into legal ones that can be presented to the judge.

Evidence in the form of testimony by an expert witness, for instance on the processing of an image to reveal important details or to match those details to an object, is accepted on the following grounds: First, the expert witness is in the unique position of being able to testify as to his or her opinion, which is presumed to be based on the expert knowledge, skill, and experience that he or she possesses. Other witnesses can only testify as to facts, not give their opinions. The expert testimony is (as noted above) intended to assist the jury in understanding the evidence in order to determine the facts at issue in the case. The basis for the testimony is expected to be an "accepted and reliable" methodology. This can mean different things at different times, and there have been recent changes in the U.S. rules of evidence. But, in brief, it means that the methodology has "substantial acceptance" in the specialized community that uses it.

The older standard for expert testimony (the *Frye* rules, named after a 1923 case) required that the science relied upon in the testimony be "generally accepted" in the relevant scientific field. This is the standard still applicable in about half of the states. The other half, and federal courts, now use the *Daubert* test, which dates from 1993. The new requirements seem to be more specific, but in fact reveal the disconnect between scientific thought and legal terminology. They require that the science has been tested, peer reviewed, and published, that it be generally accepted in the relevant community, and that it have a known error rate.

It is the last point that raises most of the problems, because in only a few situations (the *Daubert* case involved epidemiological and toxicological studies in which statistics were central) can the probability of a procedure producing the wrong answer be quantified. In imaging, for instance, it is relatively straightforward to estimate the error in dimension that may be obtained by measuring the length of a footprint or the depth of a scratch taken from an image. But what is the probability that a surveillance photograph of a perpetrator does or does not match the defendant? An endless debate about the factors involved in identification by different procedures never resolves to a simple, quantifiable, "known error rate."

There is also a potential conflict between the wording of the test and the way scientists understand the development of knowledge by the scientific method. There are actually degrees of reliability for a procedure. Cutting edge knowledge or methods are generally ones that have been peer reviewed and published, but are not yet generally and widely in use. Yet these may be of great power and represent the very finest tools that can be employed. All accepted theories start out this way. Then there is the "standard literature," which represents the knowledge and methods that the majority of working scientists would rely upon; this is also in the published journal literature, and perhaps in a few books. Finally there is the basic underlying compilation of knowledge that can be found in most established textbooks. This is taught to students, but represents an older standard that they will be expected to go beyond once they undertake their own research. Which level should be used in the courtroom? The best method is to use the techniques necessary to obtain results in each case, but be prepared to justify your decision.

Demonstrating that a methodology has been misunderstood or misapplied, or is not widely used or is not widely accepted, may invalidate the method and hence the expert testimony. It is usually enough to demonstrate that a method (e.g., for image processing) is cited and explained in the majority of textbooks on the subject to establish it as acceptable. Newer methods that have not become part of the standard curriculum may be supported by publications in peer-reviewed journals. A method used only by the expert who is testifying, which has not been published in a recognized journal and subjected to peer review, and which is not obviously based on accepted scientific methods, or which is being used for a purpose different from its originally published intent, should be rejected.

It may be the role of an expert witness to accomplish this demolition. In one case I was involved with a situation of this type. The previous testimony from a witness had demonstrated an image processing method that appeared nowhere in the literature except for one publication in a trade magazine (not a peer reviewed journal) written by the witness himself, in which there were example photographs and a cute name for the process, but no explanation of the algorithm used. Using accepted methods I could not reproduce anything like the result shown by the expert, and I had not been brought into the case in time to prepare cross-examination questions for him. I tried to explain to the jury that the method was not documented, but did not feel confident that this was understood as anything more than one elitist scientist picking at another.

So I set up my computer with the image in question and showed them a series of accepted processing methods which were unable to enhance the image to show the detail in question. This may have opened some area of doubts, but as negative evidence it wasn't entirely convincing. Finally, I showed that I could produce the same result as shown in prior testimony only by literally drawing it in with the mouse. Then I took another unrelated picture and "found" the same evidence there, again by drawing it in. That actually seemed to get the response we wanted, which was that the evidence wasn't really in the image but had been created there by the use of an inappropriate methodology. I still don't know what the original witness actually did — I doubt that he just drew it in — but I am confident that the result was an artefact of some poorly applied image processing.

The role of the jury is to decide whether or not to accept evidence, and the opinion of the expert who explains it, in deciding the question at hand. It is not necessary that they understand the entire basis for the opinion, but obviously a witness who can explain clearly and plausibly the logical and scientific basis for his or her opinion is more likely to gain credibility for the

opinion than one who simply pronounces an opinion based on lofty credentials. Teaching the jury is a role much like teaching anyone else, and those
who have had experience in making science interesting and convincing to
undergraduates will find the same skills useful in the witness box. Remember
to direct your efforts at the jury. It is pointless to try to educate the counsel
asking the questions. If it is the counsel on your side, they already know the
answers (they've asked you the questions beforehand, or perhaps you've
helped to write them). If it is the opposing counsel who is cross-examining
you, they don't care about the answers and are just trying to trick you into
some misstatement or confuse or bore the jury. Besides, as one of my friends
has suggested, most lawyers have probably selected their profession because
they couldn't cut it in a technical field anyway.

It isn't the job of the jury to understand the expert opinion on the
evidence but to decide whether or not to believe it. That means that the job
of the expert witness is to be clear and convincing, not to alienate the jury,
not to show any bias, and to respond to the questions of counsel fully, but
not to volunteer information. The usual oath "to tell the truth, the whole
truth, and nothing but the truth" is a little misleading here. It isn't the "whole
truth" — assuming there is such a thing and the expert knows it — that is
wanted. It is truthful answers to specific questions, preferably explained in a
way that will assist the side that hired you. This includes declining to answer
questions that lead you out of your area of expertise. The jury may not know
the science, but they can probably understand human nature and they will
figure out whether you really know what you are talking about and are self-
confident without being arrogant.

It does great harm to the credibility of the expert to engage in the
adversarial attitudes and roles of the opposing counsel, but it can be hard to
remain calm in the face of personal attacks on your knowledge and persistent
efforts to trick you or misquote you. My method is to take a deep breath and
remember that I'm being paid by the hour, and the longer they keep it up,
the better. A favorite line of attack on an expert witness is always "and how
much, Dr. Russ, are you being paid to come in here and defend (or convict)
this person?" My answer is that (a) I have been retained to explain and
comment on the evidence and to help the jury understand it, not to defend
or convict anyone, and (b) I am being paid about the same as the attorneys
on an hourly basis, the same rate that I charge to industrial clients who wish
my expertise. That usually ends that line of inquiry, and doesn't seem to upset
the jury. Having a high hourly consulting rate is proof of your expertise, not
something to apologize for.

Remember also that the standard of proof in a legal proceedings is far
less than that which scientists expect to encounter in their field. In a civil
trial the plaintiff's burden of proof is "by a preponderance of evidence" which

apparently translates to 51%, or even less if you imagine that only ten-twelfths of the jury must be 51% convinced. In a criminal trial the standard is higher, requiring a unanimous jury for conviction with guilt proven "beyond a reasonable doubt." I've often seen that described as perhaps 90% certain, hardly the confidence limits that a scientist would require for data before publishing a conclusion.

In many cases, particularly civil proceedings such as claims of liability, the experts are brought in long before a trial. In fact, many of these cases are settled without trial once the expert testimony has been obtained in depositions. The same thing happens occasionally in criminal trials, either because the evidence turns out to be enough to convince the accused to plead guilty, or because it turns out to be exculpatory and convinces the prosecutor to drop the charges.

As an example, in a civil case involving a large amount of damage, it was claimed that industrial X-rays of a transfer pipe routinely taken before the accident showed flaws that might have triggered the eventual rupture. Photographic prints made from the X-rays did indeed show small spots that might have been voids in the metal, but I am never comfortable examining prints because they are so much poorer in tonal range than the originals. With X-ray film, it is particularly difficult to examine the full tonal range of the negative, so when the original negatives were finally provided (which involved transatlantic shipment) I digitized them so that the tiny suspect spots could be examined. They still looked like little voids, which might have been related to the accident. But in the process of putting the negatives onto a light box so they could be photographed using a digital camera and microscope, I noticed that the spots were not in the negative but on it. They were dried water droplets from the development process, not anything in the metal at all. The "evidence" of the X-ray films was quickly dropped from the case, and no further action from me as an expert witness was needed.

Depositions are quite different from trials. The process of pretrial discovery usually requires that the opinions of the experts for each side be available to the other, sometimes in the form of written reports, but often from live depositions. During the deposition, opposition counsel will systematically try to get the expert to respond to some question with an answer that can later be taken out of context and used to seemingly contradict what is being said during the trial. It is, therefore, vital that the expert be fully prepared for the deposition, and not treat it lightly because it is taking place outside the courtroom.

Answers to questions should be as brief and to the point as possible. Don't make extended statements that may include some new idea you haven't fully explored or thought through, which may later come back to haunt you. And don't try to educate the opposing side — reserve that for the jury. Keep your answers minimal and volunteer nothing. Remember that they aren't looking for information about the facts, but about you. The goal is to find ways to make you contradict yourself or to make it possible to eliminate your evidence, for example based on a challenge to its reliability (under either Frye or Daubert rules as discussed above). Take your time, particularly when you are handed a document and asked to comment on it. Take the time to read it thoroughly, and if you don't have any useful immediate response to it, say that it would take more study to comment meaningfully. And above all, don't be led into answering questions that would stretch the boundaries of your field of expertise, which can later be used to discredit or eliminate your entire testimony.

When an expert witness is first contacted by counsel it is usually because they want some potential evidence evaluated to see if it helps their client's case. It is important at that point for the expert to see all of the evidence, not just part of it (e.g., all of the pictures, not just the ones the attorney feels will be helpful). It is the role of the expert at this stage to advise counsel on just what the evidence shows and what it does not. For the latter, it is necessary for counsel to state the "theory of the crime" (there may be several alternates) that they are seeking to prove. If some of the evidence does not support one or more of these hypotheses, it is the expert's role to point that out. This is obviously best done long before reaching trial, and the expert used in this way may never be used in the trial, if indeed one ever takes place.

When trial approaches, the information provided by the expert must be refined into questions and answers. Counsel is an advocate, interested in bringing to the jury the most favorable evidence and interpretation of that evidence. It is the role of counsel to choose those avenues of questioning and the wording of questions that he or she feels will communicate best to the jury. The role of the expert now is to help counsel understand which questions and answers are most important in this regard, and which open up avenues that are less solid and may lead to cross-examination questions that might weaken their argument. Unlike a scientific dialogue in which all sides of a question are expected to be revealed and explored, in the adversarial setting of a trial (either civil or criminal) the advocates are not trying to establish truth, but rather to convince the finders of fact — usually a jury — that the preponderance of evidence supports their side of the case.

The expert, with a background in the sciences and a different tradition of open disclosure, may feel uncomfortable with this approach. It is, however, an important and vital part of our legal system and has developed over many centuries to fill a necessary role. The expert must not feel that he or she is lying or acting against their scientific training to bring forward those parts of the evidence that are most helpful to the case, nor to answering questions on the stand with just the information requested. If there are important contradictions between the evidence as the expert understands it and the case being presented, that should be brought to the attention of the counsel who retained your services. But it is their decision as to how to present the case to the jury. They are officers of the court with legal responsibilities to the system, not the expert witness.

At the same time, it is the duty of the expert to answer truthfully those questions that are asked, either in direct testimony or cross-examination. If opposing counsel ask questions that elicit information that weakens or even contradicts the case being made, then you must provide that information and opinion.

On one occasion, I was profoundly relieved by this type of question. I had been asked to testify in defense of a (subsequently convicted) burglar, suspected of a string of house break-ins, on the subject of a paint smear found on a screwdriver in his pocket when arrested, and whether or not it matched one of the houses broken into. On the basis of microscopic images, it was clear that while the paint was superficially of the same color as that on the house, it contained different pigment particles and was not the same paint. But there was plenty of other evidence against the defendant, and certainly my own judgment about his innocence was not favorable. On cross examination, after having no success in shaking any of my testimony about the paint chip, the prosecutor finally asked, "Does the fact that in your opinion this paint does not match the house in any way indicate that the defendant is innocent of the break-in?" I was pleased to be able to undermine my entire line of testimony by the short answer "Of course not."

Confrontations between two different sets of expert opinions on the same questions of evidence are not uncommon in trial situations. Usually they represent honest differences of opinion on relatively minor points, which counsel will try to amplify in hopes of undermining the credibility of an opposing witness on the more basic points that really matter. As an example: in one case I was intensely questioned for more than an hour on my interpretation of the printed manufacturer's specifications on a video recorder, and whether the statement that it provided up to 300 points of resolution for black and white images vs. 240 points for color images applied to the particular case since the camera used was a black and white camera. Their witness had used the higher value and I had used the lower one because while the camera provided only a black and white signal, the recording format was still NTSC color. But this was entirely beside the main point of contention which was the resolution of the actual recorded image. I had actually measured that (using a procedure as described in chapter 3) at about 200 points of resolution, the difference probably being due to the quality of the tape and the cleanliness of the recording head. Any difference of opinion about the written specification was quite irrelevant to the actual matter at hand. It was very difficult to find new ways throughout the repetitive cross examination to keep making the point to the jury that the actual resolution was what mattered, and that it was much less than either specified value.

There may be other situations in which the expert witness employed by one side is using what is generally called "junk science." This is harder to deal with because it invariably involves casting aspersions on the technical qualifications of the other expert, which is always uncomfortable. But carefully pointing out contradictions in their testimony, misreadings or misunderstandings of the accepted literature, inadequate credentials, etc., is quite important. In one case an opposing expert witness had used some figures from a textbook to make a point. The figures turned out to be from my book, and they had been misused with the claim that they demonstrated one thing when in fact they did not. It was of course very embarrasing for all concerned when I was later brought to the stand as the author of that book to explain what the figures actually represented and that the earlier witness had misunderstood and, hence, misused them. A few examples of that sort (which, of course, the counsel questioning me tended to belabor) undermined much of the other witness' testimony, even on some other points where she was perhaps correct.

It works both ways, of course — you never want to find yourself testifying against the person who "wrote the book." In one case that I was involved in, an opposing witness turned out to have taken a college course that I taught, some years before. The course topic was not directly related to the issue at hand. I respected the imaging work he had done for the trial and the opinion he had formed, although I disagreed with it. But counsel, seeking to undermine his credibility, asked whether he was a former student, and what grade he had received in my course. I was obliged to answer a "C," which pretty much ended his usefulness in that trial.

To reiterate a point made earlier, it is the expert's role to provide opinions based on their scientific knowledge. It is the jury's role to decide whether or not to accept that opinion. An expert who has abused the jury by providing unfounded opinions or by not supporting them with convincing explanations and accepted facts, will also have any sound opinions devalued. Juries may not know science, but they do know people, and make their own judgments about whom to believe and trust.

This reinforces the role of the expert witness as a teacher. Presenting evidence to the jury requires many of the same skills, albeit with less ability to organize an integrated lesson plan, as teaching students. Like most students, jurors have different learning styles. Statistically, most will be visual learners. A few will be kinesthetic learners who ideally learn "hands-on" by trying things for themselves. A few will be auditory learners, who are comfortable with words as the vehicle for knowledge. And a very few will be ideational learners who are primarily interested in concepts and logic. The last two styles would fit the courtroom situation well, but few people fit into these categories. To reach the predominant visual learners, you must use images to reinforce your words and logic.

For the imaging expert this would seem to be natural: use plenty of images. It takes a little more than that to communicate an effective story that will make the evidence acceptable and relevant to their subsequent decision. I've found that the best way, when time and facilities permit, is to actually set up my computer with appropriate monitor displays for the coutroom and walk the jurors through a few examples with the actual images, showing how the important evidence was extracted from the raw image data. Of course, this has to be practiced beforehand so that (like a good classroom demonstration) it compresses into a few minutes (the average attention span of the observer) a process that may have taken hours to design and carry out in the first place. Smooth, skilled delivery along with calm, measured explanation is the basis for the "seeing is believing" response that you want from the jury. You are trying to satisfy all four of the possible learning styles in this type of presentation (even the kinesthetic learner who would like to be moving the mouse for himself but can at least identify with the step-by-step approach you are taking).

When actual demonstrations are not possible for practical reasons or constraints of time, it is still possible to capture some of the same flavor by preparing a series of printed images that work through examples step by step, which the jury can handle for themselves, that demonstrate the procedures used. This does a lot to demystify the "computer magic" that extracted the critical information from the original images.

Juries deliberate in private, so the exact influence of expert testimony on their decision process is very hard to evaluate. Generally it is believed and reported that in most cases the jury does not use expert opinion very strongly in reaching their decisions. It isn't that they don't believe the witness — a good witness can definitely explain complicated science to a lay jury and earn their trust and respect, particularly if he or she remains visibly impartial about the overall case being tried and restricts his or her answers to the matters of scientific evidence being presented. It just seems that juries don't think that experts add very much to the overall body of proof in most cases. So much for the scientific method of seeking truth! But at the same time, the information and education provided to the jury by an expert witness can provide a filter through which they view the evidence, so that without realizing that they are relying on the expert testimony it can have a powerful effect on their final decision.

Index